# An AFS Guide to Fisheries Employment

Edited by

**Tracy D. Hill**
Department of Wildlife and Fisheries Sciences
South Dakota State University

and

**Robert M. Neumann**
Department of Natural Resources Management and Engineering
University of Connecticut

American Fisheries Society, Education Section
Bethesda, Maryland

**An AFS Guide to Fisheries Employment**
is a special project of the
**Student Subsection of the Education Section**
**American Fisheries Society**



January 1996

ISBN 0-913235-98-9

# Contents

CONTENTS

# Contributors

Neil B. Armantrout (Chapter 6): U.S. Bureau of Land Management, Eugene, Oregon 97440.

Bruce A. Barton (Chapter 8): Department of Biology, University of South Dakota, Vermillion, South Dakota 57069.

Hannibal Bolton (Chapter 6): U.S. Fish and Wildlife Service, Arlington Virginia 22203.

Paul Brouha (Chapter 13): American Fisheries Society, Bethesda, Maryland 20814.

Linda Erickson-Eastwood (Chapter 5): Minnesota Department of Natural Resources, St. Paul, Minnesota 55155.

Stephen A. Flickinger (Chapter 1): Department of Fishery and Wildlife Biology, Colorado State University, Fort Collins, Colorado 80523.

Cay C. Goude (Chapter 4): U.S. Fish and Wildlife Service, Sacramento, California 95825.

Fred G. Harris (Chapter 5): North Carolina Wildlife Resources Commission, Raleigh, North Carolina 27604.

Tracy D. Hill (Coeditor): Department of Wildlife and Fisheries Sciences, South Dakota State University, Brookings, South Dakota 57007.

Gene R. Huntsman (Chapter 7): National Marine Fisheries Service Beaufort Laboratory, Beaufort, North Carolina 28516.

Robert L. Kendall (Chapter 13): American Fisheries Society, Bethesda, Maryland 20814.

Andrew J. Loftus (Chapter 10): American Sportfishing Association, Alexandria, Virginia 22314.

# CONTRIBUTORS

Donald MacDonald (Chapter 12): MacDonald Environmental Sciences, Ltd., Ladysmith, British Columbia, VOR 2EO, Canada.

F. Joseph Margraf (Chapter 9): West Virginia Cooperative Fish and Wildlife Research Unit, Morgantown, West Virginia 26506.

Bruce W. Menzel (Chapter 1): Department of Animal Ecology, Iowa State University, Ames, Iowa 50011.

Christine M. Moffitt (Chapter 4): Department of Fish and Wildlife Resources, University of Idaho, Moscow, Idaho 83844.

Brian R. Murphy (Chapter 9): Department of Fisheries and Wildlife Sciences, Virginia Tech, Blacksburg, Virginia 24061.

Richard A. Neal (Chapter 7): Southwest Fisheries Science Center, La Jolla, California 92038.

Robert M. Neumann (Coeditor): Department of Natural Resources Management and Engineering, University of Connecticut, Storrs, Connecticut 06269.

Richard L. Noble (Chapter 2): Fisheries and Wildlife Sciences, North Carolina State University, Raleigh, North Carolina 27695.

Forrest Olson (Chapter 12): CH2M HILL INC., Bellevue, Washington 98009.

Donald J. Orth (Chapter 3): Department of Fisheries and Wildlife Sciences, Virginia Tech, Blacksburg, Virginia 24061.

Nick C. Parker (Chapter 11): Texas Cooperative Fish and Wildlife Research Unit, Texas Tech University, Lubbock, Texas 79409.

Arden J. Trandahl (Chapter 11): D. C. Booth Fish Hatchery, Spearfish, South Dakota 57783.

# CONTRIBUTORS

Michael J. Van Den Avyle (Chapter 2): Georgia Cooperative Fish and Wildlife Research Unit, University of Georgia, Athens, Georgia 30602.

David W. Willis (Chapter 1): Department of Wildlife and Fisheries Sciences, South Dakota State University, Brookings, South Dakota 57007.

# Preface

During the 121st Annual Meeting of the American Fisheries Society (AFS) in Pittsburgh, Pennsylvania, the Student Subsection of the Education Section was formed. The intent of the Subsection was to promote involvement of students in Society activities and to enhance communication among students and other members within the Society. A primary interest among members during the early stages of Subsection activity development was the possibility of creating a fisheries employment handbook to aid students who are looking for jobs. Elise Irwin and Michael Allen, past Subsection presidents, were receptive of the need for the handbook and began developing ideas for content and production.

Early intentions for the handbook were to provide a detailed, comprehensive reference that included items such as telephone numbers, agency testing dates, etc. However, as ideas for handbook content developed over the past few years, members of the Subsection agreed that the focus of the fisheries handbook should be on general employment information for a number of fisheries-related careers, rather than a comprehensive, detailed reference that may have become obsolete over the years as agency-specific information such as telephone numbers, contact people, and application dates changed.

The initial selection of chapter topics was decided upon by members of the Subsection with help and cooperation from other members of the Education Section. Once specific topics were decided upon, we were faced with the task of selecting authors for chapters. We asked several fisheries professionals for their advice on who we might ask to write each chapter. We thank the following individuals for their assistance on recommending authors to write this handbook: Ira Adelman, Neil Armantrout, Ray Hubley, Donald MacDonald, Larry Olmsted, Ira Palmer, and Jack Wingate.

The objective of this handbook is to provide a source of information for those interested in a wide-variety of careers in the fisheries profession. The initial intent was to provide a reference for students seeking employment; however, non-students who are interested in other possible career opportunities may also find this handbook beneficial. As ideas for handbook content developed, we realized that in addition to providing information specific to various fisheries-related careers, handbook information should also include other professional development related topics ranging from curriculum development to building an effective resume to maintaining professionalism while involved in a fisheries career. This handbook was designed to provide general employment information for a wide-variety of career types including state, federal, provincial, and international agency employment. Aquacultural, private consulting, and academic careers in fisheries are also covered. Information regarding equal opportunities in fisheries employment is also provided in this book.

Each of the chapters in this handbook were reviewed by a review board composed of Subsection members. We believed that because this handbook was being produced by the Subsection, and because students were the primary target for the handbook, students would make excellent reviewers. Non-student fisheries professionals also contributed their time to help with the review process; their help, ideas, and guidance were invaluable. We thank the following individuals for their time and effort in reviewing this handbook: Mike Allen, Bruce Barton, Brian Blackwell, Robin Hill, Paola Ferreri, Kevin Pope, and Donovan Verrill.

Funding for this book was made possible through fund-raising activities by the Subsection at national and regional meetings. We thank all those who purchased fund-raising items and hope that they find their money was well spent. Also, we are indebted to the Education Section for their partial funding; many activities of the Subsection would not be possible without initial support from the Education Section.

Finally, we extend our gratitude to authors of this handbook. We were overwhelmed by the amount of encouragement and support given to the project by the authors. During handbook development we never forgot that, as with many AFS activities, this handbook was being produced through volunteer efforts of the authors. We gratefully acknowledge their time and commitment, and we know that their efforts will be invaluable to members looking for fisheries employment for years to come.

T. D. HILL
R. M. NEUMANN

Chapter 1

# Getting Started Early: Undergraduate Fisheries Curriculum Development and Professional Involvement

DAVID W. WILLIS, BRUCE W. MENZEL, AND STEPHEN A. FLICKINGER

Many people are surprised to learn that fisheries curricula typically include much math and physical science, as well as the expected biological sciences. However, fisheries professionals are indeed scientists, and actually need to either undertake their own research or be able to interpret and use research completed by others. People who like to hunt and fish but have little aptitude in science may be wise to choose another profession so they can afford to hunt and fish at their leisure. Indeed, increasing numbers of individuals are entering the fisheries and wildlife professions because of an interest in environmental conservation, rather than because of interest in fishing or hunting.

A **fishery** consists of three interacting components. **Habitat** is a necessary requirement for all living organisms, and includes both the living and non-living portions of the surrounding environment. **Humans** are the second component of a fishery. Humans are both users of fishes themselves and competitors for water. Many human activities, such as manufacturing or agriculture, require substantial amounts of water. Finally, fishes are part of the **biota** -- the third part of the fishery triad. Biota include both target species (e.g. sport fish) and non-target species such as prey fishes, aquatic insects, etc. **Fisheries management** is the manipulation of the three elements in a fishery to meet intended and desirable objectives. Thus, a fisheries education must prepare students to understand habitat needs and human effects, as well as the fishes themselves.

## CURRICULUM

How should a person evaluate prospective undergraduate programs in fisheries and aquatic sciences? Well, a good starting point is the professional certification program that is administered by the American Fisheries Society (AFS). After completion of a bachelor's degree, professionals can apply for certification as an Associate Fisheries Scientist if their program of study provided adequate education. Minimum requirements include at least 30 semester hours (45 quarter hours) in biological sciences, including at least four aquatics courses with at least two in fisheries; at least 15 semester hours (22 quarter hours) in physical sciences such as chemistry and physics; at least six semester hours (nine quarter hours) in mathematics and statistics; and at least six semester hours (nine quarter hours) in communications, including writing and speaking courses. These minimums at least provide a base from which to begin the selection process.

Based on the fishery triad, you should now see that your undergraduate curriculum should lead to an understanding of habitat management, the human dimension in fisheries management, and the fishes and other living organisms in the system. The goal of undergraduate education at all universities should be to provide a broad education, including humanities and sciences. No university is simply in the business of providing technical training in a narrow area. This philosophy of broad education is essential in fisheries and aquatic sciences.

Students should further select an undergraduate program based on its ability to provide training in the diverse areas of responsibility typical of entry-level fisheries positions. Graduates should have an understanding of the following basic qualifications:

- physical, chemical, and biological processes (including population, community, and ecosystem processes) in aquatic environments;
- the scientific method, experimental design, and sampling procedures;
- library research and information retrieval;
- water quality analysis and interpretation;
- identification of common aquatic invertebrates, algae, and vascular plants;
- fish ecology and biology, especially fish growth, environmental requirements, food resources, reproduction, mortality, culture, life history, movement, and distribution;

- fish anatomy and morphology;
- fish sampling methods;
- analysis of fish samples for length, weight, age, and sex;
- estimation of fish population characteristics and assessment of fishing impacts on populations;
- animal welfare concerns;
- concepts of natural resource management and related social and economic issues;
- basic statistics, word processing and computer analysis, management and summarization of data;
- oral and written technical communication, popular writing, public speaking, and public relations;
- fundamentals of budget preparation and management, and employee supervision.

Not all programs provide background in all of these areas; however, the combination of education and experience gained by graduates should include most of these qualifications.

This list of basic qualifications indicates the reasoning behind the minimum requirements for AFS certification. The influence of the fishery triad is also obvious. Students need a solid background in physical sciences and mathematics. Biological science coursework should begin with foundation courses, and then progress to upper-division specialty courses and specific fisheries courses. Communication courses are essential to employees, and wise students will prepare themselves thoroughly with a background in both verbal and written communications. Coursework involving sociology, economics, and public relations will be valuable for dealing with the human dimension of fisheries.

A geographical perspective is generally taken by most university programs, and certainly influences potential employers. Students whose intention is to work with trout fisheries in the western United States would be well advised to attend a university in that geographic area. Students that wish to eventually work on southeastern reservoirs might want to consider choosing a school in the southeastern United States. Persons with interests in marine fisheries would be better served at universities in coastal states. In our experience, inland schools that focus on freshwater fisheries tend to have weak or nonexistent marine programs. Some fisheries professionals argue, perhaps correctly, that diversity in background leads to a better scientist. However, geographical bias does exist, and students at least need to be aware of it.

## WORK EXPERIENCE

Although some undergraduate coursework will include field experience, most students that obtain fisheries employment will have obtained further work experience. Very often, this can be accomplished through summer job experiences. Some university programs have formal requirements for internships or summer work experience; others simply stress the need for this experience and help undergraduates find internships and summer aide positions. Most state and federal conservation agencies hire undergraduate students during the summer. In addition, universities with a successful graduate fisheries program commonly hire undergraduates as summer technicians to work with graduate research assistants. Undergraduate students will find these summer positions to be rewarding. They provide an opportunity to apply techniques learned in the classroom, hone your skills, and enjoy your chosen profession. In addition, students should also seek opportunities to gain experience through volunteering for various activities. Volunteer activities can take many forms. For example, many state conservation agencies have field spawning operations for fishes such as walleye, northern pike, striped bass, rainbow trout, brown trout, brook trout, or Pacific salmon. Volunteering to work at such an operation can provide valuable experience in fish culture, fisheries management, and public relations, and also provides an opportunity to interact with working professionals. Students can again look to the graduate research program at a university to provide opportunity for volunteer work experience. Students might volunteer to gain experience in fish sampling techniques, fish marking, care and handling of captive fishes, and endless other possibilities. Finally, some universities have field biological stations where a substantial amount of field experience can be gained. This experience may be gained throughout formal coursework at the field station, and through independent study.

## PROFESSIONAL INVOLVEMENT

The sooner that students become involved in professional societies, the more likely they are to gain employment and eventually contribute to the fisheries profession. For many students, undergraduate student clubs may provide the first opportunity to become professionally active. These clubs have a variety of names, and a variety of affiliations. Some are affiliated with AFS, some with The Wildlife Society, some with other professional societies, and some function effectively without affiliation. These clubs provide an important opportunity for undergraduates -- the opportunity to serve. Your first service may simply be as a committee member. However, you will then have the opportunity to progress in your responsibilities, perhaps to committee chair, and perhaps to an elective office within the club. Along the way, you will learn skills that will be essential in your professional career.

There are a few AFS chapters that are directly affiliated with university programs. Many other university programs have a formal or informal relationship with their state or local chapter of AFS. Activity in such programs will obviously be enjoyable and rewarding. However, student members of AFS need to realize that they have all the rights and responsibilities of any other member. Students are often unsure of their position in the society, but they have one vote on important issues just as do all other members. Thus, student involvement in AFS at the chapter, division, section, or society level is common and welcomed by the society as a whole. Both undergraduate and graduate students who are members of AFS are eligible for travel grants to attend meetings at most levels of the society. Each year, students can apply for a Skinner Memorial Travel Award to defray costs of travel to the annual AFS meeting. Other grants are available at the division and chapter level of AFS. Attendance at local scientific meetings can also be valuable. University department chairs or heads recognize the value of professionalism and often provide vans to take students to professional meetings that are held within a reasonable distance from campus. An active student club often can facilitate such travel.

## QUESTIONS YOU SHOULD ASK

Before making your final selection of an undergraduate program, the following list of questions should help you assess program quality.

- Does the university teach the courses required for professional certification?
- Does the content of the fisheries and aquatics courses cover the topics needed to qualify for entry-level positions?
- Are intern programs, field courses, and relationships with natural resource management agencies sufficient for students to gain practical experience?
- Are the fisheries faculty active in research and do they provide students with opportunities for research experiences and interaction with graduate students?
- Do the faculty publish research articles and maintain membership in AFS and other aquatic-oriented professional societies so they remain current on advances in the profession?
- Is there a fisheries-related student organization to provide opportunities outside of the classroom for interaction among students with common interests?
- Are there several fisheries scientists on the faculty to provide students with a broad spectrum of views?
- Are class sizes small and are there sufficient opportunities to interact with faculty on a formal and informal basis?
- What is the program's record in placing students in fisheries and aquatic positions?

Availability of these opportunities in a university program does not insure a high-quality fisheries education, but their absence is likely to result in a less-than-satisfactory education.

## FURTHER SOURCES OF INFORMATION

The information on qualifications for fisheries jobs and questions to ask about a potential undergraduate program were obtained from the AFS publication "Fisheries Programs and Related Courses at North American Colleges and Universities." This publication provides information on undergraduate and graduate, fisheries-related programs at North American colleges and universities. Mailing addresses, degrees awarded, and number of fisheries-oriented faculty are provided for each program. This useful publication is available from AFS (5410 Grosvenor Lane, Suite 110, Bethesda, MD 20814), and from the AFS Campus Network Representatives that serve at the primary university fisheries programs across North America. The annual Aquaculture Buyer's Guide, published by *Aquaculture* magazine (31 College Place, Asheville, NC 28801), provides information on schools, coursework, and degrees offered for aquaculture programs.

Chapter 2

# Building a Resume

RICHARD L. NOBLE
AND
MICHAEL J. VAN DEN AVYLE

Looking for a position? One of the most frequently encountered requests will be, "Send a resume." You'll need to respond with an advertisement of your personal and professional skills, traits, and talents that is primarily intended to convince employers to contact you for further information. A resume, usually of two pages or less, summarizes your professional capabilities in a way that allows employers to determine if you meet their needs. A resume will make a strong first impression if it is visually appealing, stresses your productivity, demonstrates a high level of literacy, and highlights your interests and abilities. Whether submitted as part of an application, handed to a potential employer during a casual encounter at a chapter meeting, or requested by a reference as a memory jogger, this little document may be the key to getting the position you really want.

Because the resume may be needed for a variety of situations, it must be versatile and concise. For preliminary screening it should be perusable in less than a minute; but it also must provide enough information to rank candidates for interviews. The resume may be reinforced with a letter of application, but it must be complete enough to stand on its own. And, for the interested employer, it must provide convenience of return contact with you.

Perhaps the best way to think of your resume is as your best representation of yourself in your absence (although you'll frequently be there with it!). Use it to summarize your abilities relative to the desired position and to sell yourself as the best person for the position. Unlike the formal application form which constrains what you can provide, the resume is **your** self-portrait, painted to emphasize your best aspects. Your resume should reflect individuality and not be suppressed by conformity to standards. However, organization for quick identification of your various attributes is essential. So, incorporate and modify the following recommendations to fit you and your needs.

## CHARACTERISTICS OF AN EFFECTIVE RESUME

Careful thought and planning are required to prepare an effective resume. Select the best material, organize it well and present it flawlessly. Doing otherwise will negate the effect of even the best professional's accomplishments. A resume will be ineffective if it is not "on target," is too long or too short, is wordy, or uses poor language, writing style, and tone.

### First Impressions Are Important

Besides being short, a resume must have an appearance of high quality and organization. Key information should be presented without pages appearing crowded or cluttered. Use of empty space, variable font or type sizes, distinct headings, and a balanced layout enhance the appearance and readability of a resume.

Choice of paper color and quality also affect the immediate visual impact of your resume. Black or dark blue ink on paper colors of white, off-white, ivory, and light grey provide an appealing contrast. Paper should be 20-50 lb bond, 100% cotton fiber, and must be standard business size (8.5 x 11 inches).

### Format Options

Proper organization will complete the immediate effect you wish to achieve. Although the resume will typically begin with personal information, the education and experience portions can follow either a chronological or functional format (or a mix of the two).

A chronological format is the most common. It is a complete listing, by year, of work-related activities within certain categories, e.g., Education, Employment History, and Professional Activities. Within each category, activities are usually listed in reverse chronological order (most recent first). The reader can quickly determine "what, when, and where" about your professional background.

A functional format summarizes experience and accomplishments in topical areas, e.g., Teaching, Research, Administration, and Practical Skills. Past experience is concisely summarized by describing major accomplishments. For example, under a Research heading, an applicant could write: "Worked total of

15 years in federal, state, and private research organizations. Successfully secured outside funding totalling $10 million, directed work of eight permanent research scientists, and published 75 articles in scientific journals." A very brief listing of recent accomplishments (past 1-3 years) would be included as supporting evidence. Hence, a functional format is most useful for applicants having a long work history.

Students at the bachelor's or master's level will probably find a chronological presentation most effective for their resumes. Doctoral students and postdoctoral associates, who have a longer professional history, should consider using a functional format, depending on the position sought.

**Write Right**

Your writing style is a reflection of you. Avoid the use of first person (i.e., "I"); use active voice and meaningful active verbs which imply "action" by you. Your resume should be flawless in grammar, spelling, and punctuation, and be free of typographical errors. Detection of even one mistake by the recipient is likely to result in rejection of your application. Review all material at least four times--by you, by a friend, by an expert (professor, English major, writer, editor, etc.), and again by you. Pick the most critical reader/editor you know!

## RESUME CONTENT

Your resume will need to be updated regularly, so present personal and professional information accordingly. Some possibilities for inclusion are provided below. Don't be overly concerned about revealing private information, e.g., your social security number, age (birth date), disabilities, etc. Selectively emphasize the positive.

**Date**

Dating your resume with month and year will help the reader determine that the information is current and allow users to easily replace an outdated version with a new one.

**Personal Data**

The employer will be interested in your personal as well as

professional fit to the position and agency. Personal data will be of particular use to personnel officers after you have been selected for interview or appointment. Consider including the following:

- name;
- social security number;
- birthplace/citizenship;
- address, present and permanent, and special instructions if a change is imminent;
- phone and FAX, at home and at work, instructions on times reachable, availability of answering machine, E-mail address, etc.;
- birth date (not age);
- sex, particularly for names like Ashley, Jamie, etc.;
- family/marital status;
- military status;
- health/disabilities.

**Career Objectives (what are you seeking?)**

Be specific, but not too narrow. When your resume is submitted with a letter of application or transmittal for a specific situation, you can refine your career statement to fit. Emphasize activities, accomplishments, and type of work responsibilities sought, not position titles. Instead of: "Desire entry level position in fisheries," use a statement such as: "Desire position involving assessment and management of fish populations, habitat, and angler characteristics, leading to district-wide management responsibility." The latter is more informative about your interests and career direction.

**Educational Experience (or Education and Training)**

List degrees earned (or expected, if imminent), institution, major field and dates. List the most relevant educational experience first, generally in reverse chronological order.

- Ph.D. - include dissertation title (unless listed as a publication) and major professor
- M.S. - include thesis title (unless listed as a publication) and major professor
- B.S./B.A. - include all institutions attended
- Associate Degree, if any

- Military education
- Correspondence courses
- Do not list high school

Academic honors (e.g., summa cum laude, dean's list, scholarship awards) may be noted as a part of each degree or presented in a separate section of the resume if justifiable. If you have achieved academically, perhaps in the B-average range or better, you may want to specify your grade point average (GPA). If your academic achievements have been particularly strong in your major, you may wish to supply your GPA in Major in addition to, or instead of, your overall GPA. Be sure to indicate the scale of GPA measurement (e.g., 3.27/4.0).

For undergraduate students and graduates of institutions with degree titles other than "Fisheries," a listing of major courses may be an appropriate addendum to the Education section. Avoid use of the term, "relevant courses," since relevancy may be in the eyes of the beholder. Courses to list here would include those in aquatic science and natural resources, communications, statistics, and computer science.

**Employment History**

The Employment History section of your resume may be alternatively entitled Employment, Employment Background, Employment Profile, Professional Background, or Career Profile. In this section, emphasize jobs and experience that best demonstrate your competence relevant to the desired position. Thus, you may wish to develop different resumes for job applications than for graduate school inquiries. Like education experience, job experience should be listed in reverse chronological order because your most recent experiences are likely to be most relevant. If not, make sure your most relevant experience is highlighted near the beginning.

But what is relevant? Demonstrate motivation, leadership, and initiative, as well as technical competence. Positions that are clearly professional in nature obviously should be included. Don't forget to consider part-time positions, internships, co-op assignments, volunteer positions and military experience. Depending on how you title this section, you might even fit in some class experiences. For undergraduates and recent graduates, jobs unrelated to professional goals may help demonstrate your work ethic, especially if they substantially contributed to paying your way through school. Simultaneously working and achieving academically can be an important attribute in the eyes of employers. But don't overburden your resume with non- professional experiences, especially if you have some professional accomplishments already listed. You might even reduce them to a summary statement. For example "Earned support for school expenses by working part-time (20-30 hr/week) for three years. Pumped gasoline, grilled hamburgers, and cleaned dormitories."

For each position, provide informative documentation including:

- dates of employment;
- position title;
- employer and address (agency, organization, department, etc.);
- brief statement of major responsibilities or accomplishments. Emphasize skills and accomplishments, rather than duties and responsibilities. For example: "Checked fish populations in 23 private ponds using seines and electrofishers and advised landowners of appropriate management procedures" is more impressive than "Collected fish with seines and electroshocker and recorded data";
- names of supervisors (optional);
- do not include reasons for leaving positions.

**Other Accomplishments and Skills**

Depending on the amount and variety of qualifications that remain to be expressed, this section is highly flexible in name, content, and organization. If the items to be included are numerous, you may want to use subheadings; conversely, some categories might be placed under headings of their own if your accomplishments are extensive enough (e.g., Publications and Presentations). Again, organize this material to highlight your most significant accomplishments relevant to the position sought.

*Memberships in professional groups*. Organizations with which you affiliate illustrate professionalism as well as your values and philosophies. What you list here will vary, depending on whether you are an undergraduate or new Ph.D. Consider student organizations, professional societies, and conservation organizations for inclusion. If you include honoraries such as Alpha Zeta, Sigma Xi and Phi Sigma, provide clarification of their purpose. Emphasize officerships, committee assignments, and chair responsibilities. You may prefer to highlight accomplishments that illustrate your productivity and enthusiasm rather than to present a long list. For example, "Belonged to nine student organizations and clubs as an undergraduate. As President of Wildlife Society Chapter raised $10,000 to fund student travel. Served on committees in other organizations (e.g., Tri Beta, Students for Environment) aimed at educating the public about environmental issues."

*Awards and special recognition*. Most academic recognition should be reported as Educational Experience, so this section can be restricted to non-academic awards. Likewise, any awards associated with publications or scientific paper presentations may be built into sections described below. If you have accomplishments that fit here, don't be timid about expressing them. You probably won't want to list items prior to entering college unless they are quite unusual and highly pertinent to the profession.

*Publications*. List all published materials in reverse chronological order. These may include reports, magazine articles, brochures, journal articles, thesis and dissertation. Include, and appropriately denote, papers that are "accepted" or "in press." If your list is long, you will want to separate the latter under a subheading. Listing of papers that are "in preparation" probably will be more indicative of procrastination than productivity. Work that forms the basis for future publications should be documented elsewhere in the resume. List all publications by year; include all authors, full title, and journal/publisher. You may wish to designate award-winning papers, even if you've done so under a previous heading.

*Presentations at professional meetings*. Follow the format used for publications above, noting name of meeting and location. Be sure to list all authors, presenter, and whether the paper was invited. Note any awards received.

*Grantsmanship/grants and contracts*. Applicants for positions that will require grantsmanship should list grant proposals by year, title, funding source and amount. Even if you successfully solicited only supplemental support for conduct of a research project or summer study program, it will demonstrate your initiative, e.g., "Obtained assistance of 12 fishing club members in collection of fish for tagging."

*Special skills and certifications*. Working in field studies, especially in the aquatic environment, requires many unique skills. Any training you have received, e.g., SCUBA, CPR, electronics, and navigation, should be noted. This may also be the place to elaborate about quantitative and computer skills.

*Interests, activities and hobbies*. This topic typically becomes less important for advanced professionals, but for undergraduate internships and entrance level positions, you may wish to include it. For example, an avid offshore angler is likely to know how to operate boats, be familiar with navigation and relate to the recreational fishing public; an avid reader of non-fiction may be particularly effective at keeping up with the scientific literature.

**Professional References**

By this point, you should have presented yourself to the best of your ability. Corroboration of your credentials is equally essential. This is done by people who know you and your capabilities--your references. Most guides on resume preparation recommend inclusion only of the statement, "References furnished on request." However, we believe that references should be routinely included. The fisheries profession is relatively small and characterized by a high degree of networking among professionals, even over a wide geographic area. "Who knows you" can be as influential on employer response as what your references actually say about you, so listing your references is generally a good strategy. Listing references also helps speed the application process by allowing the employer to contact the references without additional correspondence. If you list references, put them on a separate page at the end of the resume so you can easily tailor the list to the position you are pursuing, or leave them off completely when appropriate.

Choose references best-suited for the position sought. Consider the nature of interaction you had with the reference, the likelihood that the reference will provide a positive recommendation, and the reputation of the reference in the eyes of the potential employer. Professors familiar with your academic abilities and research potential are usually best for graduate school applications. Past supervisors, including graduate students, even supervisors of volunteer work, may be most effective for job applications. References should be able to compare your characteristics with those of a broader sample of students. Do not use classmates or family friends as references.

Your list of references should be short, usually at least three and no more than five individuals. Provide full names, position title, affiliation, address, phone, FAX, and e-mail address. Be sure this information is accurate! Get permission of each reference, and, while doing so, have them check the accuracy of your information regarding them. Then, subsequently, provide them with a copy of the resume.

## SUMMARY AND RECOMMENDATIONS

An effective resume starts with competitive credentials. Experiences of all types (see Chapter 1) not only hone one's capabilities, but help focus one's career goals. Elaboration of credentials in the resume--your best representation of you in your absence--then becomes the vital step.

Recognize, however, that presentation of your resume is also important. Like your resume, your letters of transmittal, or even a transmittal in person, should reflect a confident, knowledgeable individual who is concerned with accuracy and detail. Use these contacts not only to deliver the resume, but to introduce (reintroduce) yourself, confirm your genuine interest (which can be reinforced by a few intelligent questions about the position/employer), and offer to meet for further discussion or interview. Then follow up as appropriate by phone or letter.

You may want to explore additional options for resume content and organization. Further assistance with resume preparation is available for undergraduate and graduate students from most university placement offices. Commercial printing/copying companies located near university campuses can assist with stationery, print style, and resume layout. A substantial amount of reference materials, including those listed below, can provide further suggestions. But, most of all, seek critical review by classmates, your adviser, other professors and supervisors. Then, take the spectrum of advice and tailor a resume that fits **you**.

## SUGGESTIONS FOR FURTHER READING

McDaniel, C. 1990. Developing a professional vita or resume. Garrett Park Press, Garrett Park, Maryland.

Krannich, R.L., and C.R. Krannich. 1992. Dynamite resumes. Impact Publications, Woodbridge, Maryland.

Lauther, H. 1990. The resume reference book. McFarland & Company, Inc., Jefferson, North Carolina.

## Chapter 3

# Finding a Graduate Program in Fisheries

DONALD J. ORTH AND IRA R. ADELMAN

Decisions that you make about graduate education are more important than decisions about your first job. Whereas young fisheries professionals may change jobs several times before landing that "ideal" position, for most fisheries students the master's degree is the last degree they earn, and the selection of a graduate program is made only once in a career. This chapter is organized in a question-answer format to help students make this once-in-a-career decision of the "best" program for their interests. There are four sections entitled "Before You Apply," "Exams," "The Application and Acceptance Process," and "Graduate School." We have concentrated on issues related to the decision process rather than attempting to define the strengths of specific programs in fisheries. Students should realize that the reputations of graduate programs are dynamic and the best way to discover suitable programs for you is to conduct your own personal search. But before you do, the first question is "Should I consider graduate school?"

## BEFORE YOU APPLY

### Should I Consider Graduate School?

Yes, because if you graduate with only a bachelor's degree, you are at a competitive disadvantage. You will probably find yourself competing with graduates with master's and perhaps even Ph.D. degrees for jobs. Some technician positions are largely filled by B.S. or even associate of arts degree recipients, but these are relatively scarce and will not permit you to advance to other positions in some agencies. Aside from the competitive advantages of having an advanced degree, your qualifications and capabilities as a fisheries professional can be greatly enhanced through graduate education. Fisheries science is a broad field that draws very heavily from other natural and social sciences. Graduate education will expand your depth of knowledge in fisheries and related fields. Consequently, in order to be competitive for most entry-level positions in state/provincial/federal agencies and industry/consulting, a master's degree is a definite advantage, although not always a requirement.

The decision to extend your graduate education to the Ph.D. level should be based on your career goals. Most research scientist positions and academic positions require a Ph.D. but a relatively smaller number of Ph.D.s are hired for management positions. There is some suspicion, though undocumented, that there is a bias against hiring individuals with Ph.D.s for management positions because they may be "overqualified" or in search of academic positions.

**Should I Work for a Year or So Before Applying to Graduate School?**

If you have any trouble defining your career goals and reasons for going to graduate school, or if your motivation is not strong, you should delay applying for graduate school and obtain work experience in fisheries. A one- or two-year stint as an intern or technician will give you time to interact with other fisheries professionals and decide what type of work is most meaningful to you. The work experience will make you more, not less, competitive for the graduate program of your choice.

**How Do I Learn About Schools to Which I Should Apply?**

Start with a consideration of the career goals that you have established. This will help to narrow down an otherwise long list of possible graduate programs. The American Fisheries Society (AFS) publication "Fisheries Programs and Related Courses at North American Colleges and Universities" lists 130 master's programs and 80 doctoral programs. This guide provides a complete list and is a good starting point. Some of the larger fisheries and aquatics graduate programs are listed in Box 1. You can also develop a list by checking ads in *Fisheries*, *The Job Seeker*, and fisheries department bulletin boards where announcements of assistantships often appear. There are several ways to gather more information about these schools.

1. The Peterson's Guide to Graduate Programs in the Biological and Agricultural Sciences is an annual publication that is a good place to start in compiling your list of potential

Box 1. **Selected Fisheries Graduate Programs in North America**

Auburn University
Clemson University
Colorado State University
Cornell University
Humboldt State University
Iowa State University
Louisiana State University
McGill University
Michigan State University
Mississippi State University
Montana State University
North Carolina State University
New Mexico State University
Oregon State University
Ohio State University
Oklahoma State University
Pennsylvania State University
Purdue University
Simon Fraser University
South Dakota State University
Southern Illinois University
State University of New York - Syracuse
Tennessee Tech University
University of Alberta
University of Alaska - Fairbanks
University of Arizona
University of British Columbia
University of California - Davis
University of Florida
University of Georgia
University of Guelph
University of Idaho
University of Maine
University of Maryland
University of Massachusetts
University of Michigan
University of Missouri - Columbia
University of Minnesota
University of Nebraska - Lincoln
University of New Brunswick
University of Tennessee - Knoxville
University of Toronto
University of Vermont
University of Waterloo
University of Washington
University of Wisconsin - Madison
University of Wisconsin - Stevens Point
University of Wyoming
Utah State University
Virginia Polytechnic Institute and State University
West Virginia University

schools. It will also provide you some useful data for comparing programs. Fisheries programs are listed in the Natural Resources Sciences section (Section 20) under Fish, Game, and Wildlife Management. The descriptions here will list facts, such as degrees offered, number of faculty, degree and entrance requirements, percent acceptance, number of degrees conferred, number of women and minority students enrolled, application fees and deadlines, tuition, availability of assistantships, and faculty research interests. The description will also include a faculty contact for graduate program information.

2. To obtain further information on graduate programs, send a postcard or letter requesting an application, a graduate bulletin, and financial aid information to the graduate program contact. Later you may request a detailed description of the faculty research interests and educational offerings for the programs you are most interested in. Brochures are normally available from most major programs and these will help you identify those whose interests are most closely aligned with yours. Your initial contacts via letters are much cheaper than the average cost of an application. The letter is also your best advertisement for what you have to offer; don't be shy about including your grades and experience. Spend the time to write an informative, yet concise, letter and include a resumé with the letter. Expect to get a response that provides some indication of the likelihood of acceptance and the availability of assistantship support in the next academic year.

3. College and university catalogs are also useful, but the collections available at most libraries are notoriously out of date and incomplete. The graduate school on your campus may also maintain a collection of catalogs.

4. Familiarize yourself with journals describing current research in fisheries (e.g., *North American Journal of Fisheries Management*, *Transactions of the American Fisheries Society*, *Progressive Fish-Culturist* and many others) and note who is

publishing in the areas that interest you and where they are teaching.

The whole process of developing your information base may take three or more months. Therefore, **begin the process at least one year before you expect to enroll in graduate school**.

## Should I Only Consider Named Programs in Fisheries?

At the master's level it is more important that you select a program based on overall quality of the educational program and opportunities. At the doctoral level you should place more emphasis on selecting a major advisor than a program. In either case you will find excellent programs within departments of biology or zoology, or as part of natural resources or forestry departments, schools, or colleges. Of the 130 master's programs in North America, only 19 have "fisheries" as part of the program title. Therefore, excluding all but named programs in fisheries will limit your options significantly. The advantages of programs that are listed as departments of fisheries, or combined fisheries and wildlife, may be more faculty with fisheries interests, more course offerings, and greater breadth of fisheries subdisciplines. However, rather than consider the name of the program, evaluate programs based on their faculty, course offerings, number of graduate students, facilities, and whether the program offers sufficient courses for professional certification as a fisheries scientist by the American Fisheries Society. The minimum course standard for certification consists of: (1) four courses in fisheries and aquatic sciences, two of which must be directly related to fisheries science; (2) thirty semester hours of biological sciences, which includes fisheries and aquatic science courses; (3) fifteen semester hours of physical sciences; (4) six semester hours of mathematics and statistics; and (5) six semester hours of communications.

## How Do I Evaluate the Suitability of a Graduate Program or Major Advisor?

The AFS publication "Fisheries Programs and Related Courses at North American Colleges and Universities" lists a series of questions you should ask about a prospective program. These questions were primarily developed for consideration of undergraduate programs, but most are also pertinent in considering graduate programs. An expanded and modified list of questions is given in Box 2. Your first consideration in selecting a major advisor is to seek one whose area of interest is more or less compatible with fisheries sub-specialities in which you are interested. This is of utmost importance if you are seeking an advisor for a Ph.D. program. You should ask for lists of the potential advisor's recent publications and thesis titles of previous students, and where his or her students have been employed. This information is frequently available along with application materials.

There is no substitute for visiting the campus and meeting the professor to see how well you are likely to get along. Be sure to ask current graduate students about their impression of your potential advisor. If you can't visit the campus, then a phone call is the next best alternative. It is likely that if a professor is interested in accepting you as a student, she or he will phone you and possibly ask you to come in for an interview. Finally, if you know fisheries professionals from academic institutions or agencies, seek their advice on suitable programs and advisors, understanding that the scope of one individual's knowledge of such matters may be quite limited and perhaps biased.

The relationship between student and major professor is one of the keys to successful graduate education. Because each professor has a unique mentoring style, it is critical that you meet your prospective major professor, learn about his or her mentoring style, and are comfortable with the mentoring approach. Former graduate students are your best source of information.

## EXAMS

### What Exams Will I Need to Take?

Application materials from prospective programs will tell you what exams they require. Most fisheries graduate programs require the Graduate Record Exam (GRE) general test. Some require the subject test in biology. You should definitely plan to take the General Test. If you are uncertain as to where you will apply, take the Subject Test as well. That way, if you subsequently apply to a school that requires the Subject Test, consideration of your application will not be delayed. You will be asked when you take the exam to list the

Box 2. **Questions to ask professors and students about prospective programs and advisors.**

*Questions for Professors*

- Does the university teach the courses required for professional certification?
- Do the faculty publish research articles and maintain memberships in professional societies?
- Are there several fisheries scientists on the faculty to provide students a broad spectrum of views?
- What is the program's record in placing students in professional positions?
- How accessible are the faculty to students?
- Is there support to send students to professional meetings and publish research articles?
- Is grant/contract support adequate to permit completion of thesis research in a reasonable time frame?
- What is the expected time for completion of my degree?
- Are graduate courses up-to-date, relevant to your interests and taught by well-known authorities?
- What are the demographic characteristics of the graduate student body or will I fit in?
- How many graduate students do you advise and how much time is available to advise me?
- Is your door open to students or by appointment only?
- Will I have to make up any deficiencies in coursework?
- What kind of space and lab/field equipment will be available to me?

*Questions for Graduate Students*

- Are graduate courses and seminars in your specialty offered regularly?
- To what degree are students treated as colleagues?
- What is the cost of living in the area?
- Are stipends adequate to cover expenses?
- Is there adequate public transportation?
- Are facilities for graduate education adequate?
- What is the potential major advisor's reputation with graduate students?
- Are their adequate opportunities for professional interactions on campus?
- What kind of approach do faculty take toward graduate education -- laissez faire or involved?
- What do graduate students here complain about?
- What do graduate students here like?

universities to which you want the scores sent, but you can request additional locations at a later date.

The Educational Testing Service (ETS) is gradually implementing a computer-based testing program. Currently you may take the General Test by the traditional paper and pencil method or by computer. The questions and format differ between the two exams so that the computer-based test is not just a computer version of the paper and pencil test. It is called the Computer Adaptive Test, and questions vary depending on how previous questions were answered. Scores for the pencil and paper and computer versions are interchangeable. By 1997, the computer-based test will be the only format available. At present, the Subject Test is only available in the pencil and paper form, but ETS plans to introduce computer-based versions in 1995.

**Where Do I Get Information on Taking Exams?**

Information can be obtained from Educational Testing Service, P.O. Box 6000, Princeton, NJ 08541-6000, but you may have an information center on campus at your current school that can provide all of the information you need. Most major universities have such information.

**When Should I Take the GRE?**

You should take the exam early enough to have scores sent to prospective universities at least nine months prior to your desired date of enrollment, i.e. by January 1 if you are seeking fall admission. The pencil and paper General Test will be offered five times between October and June, but as the computer version is implemented the frequency of offering the pencil and paper test will be reduced. The Subject Tests will be offered four times between October and April on the same day as the General Test. The computer-based tests are available throughout the year from three to five days per week, depending on the testing center site. Scores from the pencil and paper test are available and sent to the requested universities about six weeks after the exam. Scores from the computer test are provided to the test taker immediately upon completing the exam and are sent to the requested universities in 10-15 days.

### Will It Help Me to Study for or Take a Course to Prepare for the GRE?

Preparing for the test will help you to score better although it is difficult to provide an estimate of how much. There are a variety of test preparation books and computer programs that you can usually find at a college book store. Some major universities offer preparatory courses and there are commercial courses available. If nothing else, be sure to familiarize yourself with the test format so valuable time is not wasted during the exam.

## THE APPLICATION AND ACCEPTANCE PROCESS

### When Should I Apply for Graduate School?

Assuming you are applying for fall semester admission, you should have all of your application materials including letters of recommendation submitted by December or January, prior to the fall semester. Because there is competition among universities for outstanding students, universities attempt to make decisions about fellowships and other awards as early as possible, in part to beat the competition. Therefore, to give yourself maximum opportunity to be considered for all awards, early application is important. If you apply beyond the deadline for award nominations, all is not lost. Many university programs do not make their final decisions about admission, including research and teaching assistantships, until March, April, or later, so if your application is in by then you will be considered, but perhaps with fewer slots available than were open to earlier applicants.

You may wonder if it is advantageous or disadvantageous to apply for admission at times other than the typical fall semester. Generally there are fewer opportunities for admission at these other times but spaces are sometimes available. You should contact prospective programs about their particular situation. Sometimes you have nothing to lose because you can reactivate your application for a subsequent admission time if you are rejected because of lack of available space or advisors; with some universities you will have to pay an additional application fee.

## What Kinds of Credentials Will I Need to Submit?

You should receive an application packet with instructions from the graduate program contact. Normally you will have to submit:

- signed application form;
- application for financial assistance;
- official transcripts for all previous college courses.;
- official GRE scores from Educational Testing Service (Princeton, New Jersey);
- two or more reference letters or reference evaluation forms;
- application essay or personal statement (required by some programs);
- copy of resumé, copy of MS thesis, and reprints of publications (optional);
- name of professor you intend to study with or have contacted (optional).

In selecting references choose those who know and like you and can comment on your academic potential. Professors from advanced courses or independent studies or undergraduate research projects are usually good bets. References from supervisors in summer jobs are helpful, but don't overdo it -- one is enough. Provide a resumé to each referee, explain to them what type of position you are applying for, and inform them of the deadline for mailing. Fill out your part of the reference form, and include a self-addressed stamped envelope for their convenience. It is also wise to send a reminder to all referees before the deadline. Discuss the confidentiality of the letter with each writer; waiving the right to see the letter will usually increase its credibility. If you do not plan to attend graduate school immediately after graduate school, request that letters of recommendation be placed in a credentials file at your college or university until you begin the application process.

## How Can I Make My Application Stand Out?

This is a question that you need to ask early in your undergraduate career (See Chapter 1). Most graduate programs accept only students from the top half or 75th percentile of graduating seniors. Therefore, all the applicants will have good grades and exam scores. If your grade point average is substantially below 3.0 (i.e., a B average), you will have a difficult time being accepted in most programs unless other qualifications are outstanding. However, grade point averages are not the sole criterion for acceptance. What separates the successful applicants are other indicators of internal motivation for a career in fisheries. Reviewers will evaluate the rigor of courses, course loads, and the reputation of your undergraduate institution. There is no formula for an outstanding application because motivators and opportunities vary among individuals. Generally, reviewers are also looking at what the candidate did outside of the required courses for graduation. Details that can make the difference in acceptance and rejection include: undergraduate research project performance, leadership roles in campus organizations, work experience with progressive increase in responsibilities, internships, courses taken, honors received, society memberships and activities, writing skill, personality, maturity, and field, laboratory, computer, and statistics experience.

The application essay (or personal statement) is one of the few ways you have to convince the admissions committee of your writing ability, motivation, and commitment to a fisheries career. Be sure to express your reasons for desiring entry into that particular graduate school and your unique attributes that will contribute to the program. The essay should be critiqued by your advisor and the final copy should be neatly typed.

**How Long Will It Take for a Decision to be Made?**

When all of your application materials have been received, your application file will be sent to a department faculty committee which makes a recommendation on acceptance. Applications may be processed on a rolling basis or as a batch and the length of time for universities to make a decision on your application is highly variable. If you are a particularly outstanding applicant, you may be contacted very shortly after a program receives your application. More typically, you will hear from prospective programs between February and late April about admission the following fall. If a university is particularly interested in you, you are likely to be contacted a number of times including an offer for you to visit the campus before a final decision is made. Sometimes the prospective program will help to pay for the cost of your visit to the campus if you are a "hot prospect." Contact the graduate school or the department contact to learn of the status of your application if you do not hear anything within three months.

**Should I Request an Interview, and How Should I Conduct Myself?**

A visit may be expensive but it can be an effective way to improve your chances for acceptance or to help you make your decision on where you want to go. Normally, an interview is appropriate after you have applied and it looks like there is a reasonable chance for your acceptance and funding. If your potential advisor has not already invited you to visit, request an interview. The interview should answer questions remaining in your mind and the professor's. To make the interview work for you, be prepared with questions (see Box 2), know what you want to do in your career and be able to express it concisely, appear interested, and look good (see interview tips in Box 3). Ask to visit with current graduate students because they can answer some of your questions better (see Box 2).

## GRADUATE SCHOOL

**What Types of Financial Support are Available?**

Universities offer assistantships and fellowships to support graduate study. Assistantships require you to work, usually about 20 hours per week, for financial support. Teaching assistantships may involve delivering lectures, correcting classwork, grading papers, counseling students, and supervising laboratory groups. Research assistantships usually require that you assist in the research activities of a faculty member. Most fisheries graduate students work as research assistants rather than teaching assistants. Often work on the research assistantship contributes directly to your master's thesis or doctoral dissertation requirements; this is the most desirable funding situation. Fellowships are more competitive and generally do not require work apart from that associated with your degree requirements.

A survey of fisheries programs in 1992-93 found that the mean annual stipend for research assistantships (RA) was $9,815 for master's and $11,116 for doctoral students (Table 1). There was considerable variation among fisheries programs in stipends. Most

Box 3. **Interviewing tips.**

<u>Before the interview</u>

1. Learn as much about the position and expected qualifications as possible. In addition to the formal written description from the program, contact past graduate students and others familiar with the university and environment.

2. Obtain some background information about the university, its programs and faculty.

3. Learn who the faculty are. What are their backgrounds?

4. Develop a formal list of skills that you bring to the program. This will help you to articulate your skills in face-to-face interviews.

5. Describe your personal strengths and weaknesses. This soul-searching is difficult, but you can expect these questions. Professors are impressed with students who know themselves well. You'd be surprised how many people don't know who they are.

6. Re-examine your short- and long-range career goals. Does this graduate program's offerings fit well into your goals? Always expect a question about your career goals and why this program attracts you. Rehearse your answer with a close friend. Does your answer sound sincere or contrived?

7. Prepare questions that you have about the position, university, benefits, surrounding areas, etc., and write them down. Develop questions that will indicate your knowledge of the position or discipline.

8. Develop a list of expected questions that may be asked and answer them.

<u>At the interview</u>

1. Dress conservatively. Dress better than you expect to during a normal work day.

2. Be yourself (otherwise your other self might get hired instead of you).

3. Be prepared to initiate and maintain the interview if it is unstructured. Normally it is unstructured at universities. Prepare your questions ahead of time (#7 above) but ask them extemporaneously. Questions should appear spontaneous and unrehearsed. It is okay to do most of the talking (that's expected), but don't wander aimlessly. Look to the interviewer to direct conversation. If they don't, suggest topics that you feel should be covered.

4. Have extra copies of resumé/vita and reprints for interested persons.

5. Take notes, when appropriate; however, in most cases it may inhibit communication.

6. Be prepared to discuss starting dates in certain terms.

Table 1. Summary of annual stipend, tuition and fees (U.S. dollars) survey conducted in 1993 for fisheries graduate programs in North America.

| | Percentile | | | | | | | |
|---|---|---|---|---|---|---|---|---|
| | 20 | 30 | 40 | 50 | 60 | 70 | 80 | 90 |
| *Stipend* | | | | | | | | |
| MS (RA) | 8423 | 8990 | 9322 | 9700 | 10150 | 10680 | 11460 | 13120 |
| PhD (RA) | 9824 | 10300 | 10590 | 10920 | 11650 | 12050 | 12600 | 13460 |
| Fees | 0 | 335 | 577 | 725 | 1200 | 1518 | 2049 | 2523 |
| *Net compensation* | | | | | | | | |
| MS (RA) | 7007 | 7724 | 8494 | 8847 | 9422 | 9900 | 10280 | 11470 |
| PhD (RA) | 9013 | 9372 | 9941 | 10130 | 10500 | 10800 | 11390 | 12420 |

programs waived some or all of the fees; the required fees averaged $1,115 per year. Therefore, the net compensation after fees are paid is a better way to compare compensation for graduate students. Net compensation averaged $8,686 for master's and $10,120 for doctoral students. Although net compensation may be a better statistic for comparing programs, there are other inequities that you need to consider in selecting graduate programs. For example, some graduate programs provide excellent paid health benefits while others provide minimal benefits. Moreover, the cost of living varies widely among locations. The graduate stipend and fee statistics provided here are dynamic and have a more complicated structure at many institutions. Expect them to increase each year (perhaps 3-6%). Also, higher stipend levels are available at 58% of the fisheries programs surveyed; the higher stipend levels normally depend on grants and contracts.

College work-study and Stafford student loans and supplemental loans are available to support your graduate education. However, most fisheries graduate students are able to find support on research or teaching assistantships so they do not apply for loans or work-study. It is extremely difficult to pursue a graduate degree in fisheries as an unfunded student working part-time and we do not recommend it.

## What is Graduate School Like and How Does It Differ from Undergraduate Programs?

Graduate programs are typically more flexible and variable than undergraduate programs and thus demand more initiative and self-reliance for the student to be successful. The main goal of the undergraduate is to successfully complete a full load of courses and to participate, if possible, in special projects and other non-classroom, professionally-related activities. For the graduate student, special projects--typically the thesis, and the non-classroom professional activities are more time-consuming and perhaps more important than the formal courses. Much of the learning occurs informally through discussions with fellow graduate students and faculty, attendance at professional meetings, and through independent reading. Grades in courses are less important because students admitted to graduate programs have already proven that they can do well in formal courses. However, grades are still used as basis for academic probation and removal from many programs. Different professors have different styles in working with graduate students with extremes being, "I'm here if you need me, but otherwise you're on your own" to "Check with me before you do anything new, and here's what I want you to do." You may be more comfortable with styles tending toward one extreme or the other and you should consider those styles when selecting an advisor.

## How Long Will It Take Me to Get a Master's or Doctorate?

Expect to spend two to three years in pursuit of your master's degree and three to six years beyond the master's for a doctorate. The length of time depends more on the nature of the field or lab studies, funding support, logistics, individual motivation, and acts of God than on the specific program requirements.

## INFORMATION SOURCES

Fisheries Programs and Related Courses at North American Colleges and Universities. 1990. Available from: American Fisheries Society, 5410 Grosvenor Lane, Bethesda, Maryland 20814-2199.

Chapter 4

# Equal Opportunity in Fisheries Employment

CHRISTINE M. MOFFITT AND CAY C. GOUDE

## THE NEED FOR DIVERSITY IN PROFESSIONALS

The future of any organization or institution lies in its ability to adapt to the changing needs of society. For years the stereotypical fishery biologist was one of a white male, and historically the fishery profession was centered around sport and commercial fish interests. Women or racial minorities in the profession were unusual before the 1970s.

Changes in the scope of the fisheries discipline and in societal values are occurring rapidly. Our society has legal interests and responsibilities beyond traditional consumptive resources to that of protection of seriously declining non-game species and protection of entire ecosystems, because of human overpopulation, overharvest, introduction of exotic species, and generally deteriorated environmental conditions. This new emphasis demands an expanded approach to basic science and management issues, and broadens the funding base for work in fisheries from that based solely on sport and commercial fish to one based on understanding and protecting natural resources, water quality, biodiversity, and multidisciplinary studies including economics, law, and history.

To address global problems and issues, it is important that the profession represent the diversity of peoples associated with these issues, particularly urban issues. Not only is it inevitable that the "future" student will be one of color and equal gender representation, but it is essential that all elements of our democracy are represented and educated in the issues of natural-resource management.

## MEANING OF EQUAL OPPORTUNITY

Equal opportunity assures a level playing field, giving all candidates a chance in recruitment and employment processes. Equal opportunity differs from affirmative action. In affirmative action the minority candidate is sought out and can be chosen over an equally qualified non-minority candidate, providing that the purpose is to break down old patterns of segregation and hierarchy and to open employment opportunities which had been traditionally closed. There are cases where affirmative action is warranted, but such actions have been met sometimes with protest from those passed over. The mechanisms have been often supported by the courts who have ruled that unless positive action is undertaken to overcome the effects of systemic institutional forms of exclusion and discrimination, a benign neutrality in employment will tend to perpetuate the status quo indefinitely.

The gender and racial composition of both science educators and students in most institutions of higher learning is lagging behind that of the rest of our society. Under represented racial minorities enrolled in science include African-Americans, Hispanics and American Indians. Asian-Americans are not considered under represented in sciences, as they comprise 2.9% of the general population but 6.9% of the Ph.D. scientists and engineers (Fox 1993). In 1992, African Americans, approximately 12% of the U.S. population, earned only about 1.2% of the new Ph.D.s in science and engineering. Unless trends such as this in the training grounds for our future scientists are reversed, it will be difficult for adequate numbers of racial role models to be available. Therefore, reaching out to female and minority students at an earlier age is critical to gain interest potential scientists in the aquatic natural resources profession.

## LEGAL REQUIREMENTS

In the U.S., federally funded agencies and state agencies or institutions receiving federal funding are under legal jurisdiction of federal statutes. There is a federally mandated policy of nondiscrimination on the basis of race, color, national origin, religion, sex, age, disability, or status as a Vietnam-era veteran for all such institutions. Discrimination is prohibited by Titles VI and VII of the Civil Rights Act of 1964, Title IX of the Educational Amendments of 1972, Sections 503 and 504 of the Rehabilitation Act of 1973, the Vietnam Era Veteran's Readjustment Assistance Act of 1974, the Age Discrimination Act of 1975, the Age Discrimination in Employment Act Amendments of 1978, the Americans with Disabilities Act of 1990, the Civil Rights Act of 1991, and other state and federal laws and regulations. Furthermore, Executive Order 11375 amends

Executive Order 11246 to specifically list discrimination based on sex among prohibited discrimination in employment by federal contractors or subcontractors who employ 50 or more persons with $50,000 or more in contractual obligations.

## RIGHTS OF APPLICANTS IN THE RECRUITMENT, SEARCH, AND INTERVIEW PROCESS

A job applicant has rights. Each job vacancy should be advertised, and the vacancy should have a job description, detailing the required and optional qualifications of the job. If the job applicant has met the required aspects of the job, he/she has a right to be examined further for his/her qualifications. The hiring institution must inform the applicants of the process they will use to screen candidates.

For a vacancy, often a search committee is appointed. The committee should be representative of the broadest portion of the population interacting with the applicant. The interview process should define the qualities that are essential and required for the job and those that are optional and desired by the employer. Each applicant should be asked the same interview questions. Questions about personal life are not considered valid and if a candidate's personal life is discussed, he/she has a right of grievance. Age is not appropriate to the discussion, although past experience is relevant.

## OPTIONS IF VIOLATIONS ARE SUSPECTED

Most agencies or companies have human-resource officers. These are the first people to contact if anyone believes that the law or procedures have been inappropriate. If there are a number of people involved in a complaint, there are class-action procedures that can be used. Inexpensive legal counsel is available on most college campuses, or state bar associations through Legal Aid Associations. The Office of Federal Contract Compliance Programs in the Department of Labor is an appropriate channel for grievances against any federal contractors.

If someone is hired that is clearly not as qualified as another candidate, the candidate or other parties can request a review of the hiring process. Any conversations or other procedures that are used in a search or interview process that are considered as potential violations of the candidate's rights must be documented through phone logs, summaries of meetings, or any memos.

## DISCRIMINATION IN SALARY OR PROMOTION

Title VII makes it unlawful "to discriminate against any individual with respect to his compensation, terms, conditions or privileges of employment, because of such individual's race, color, religion, sex or national origin". This law has provided the basis for suits of gender equity in disputes of pay and promotion. If any such discrimination occurs, there are legal avenues and precedence for settlements. The "equal pay for equal work" concept is one that is being tested in several states and has resulted in review of many pay scales.

Sexual harassment can be a factor not only in the work place environment, but in the options for promotion. Educational institutions have grievance procedures to handle sex discrimination required by Title IX, and must conduct self-evaluations of these programs. Sexual harassment is explicitly forbidden in most work places. The legal basis of sexual harassment claims is continuing to develop throughout the nation.

## OTHER SOURCES OF INFORMATION AND ASSISTANCE IN CASES OF DISCRIMINATION

The American Civil Liberties Union (ACLU) is an active private organization representing the interests and rights of all workers and citizens. They have intervened on many occasions on behalf of people whose civil rights were violated. They have regional offices throughout the U.S. and phone numbers are available in most telephone directories.

## PREPARING FOR THE FUTURE

Fishery and aquatic natural resource professions are undergoing rapid changes within society. To be able to respond to these changes, we must have a profession that represents the gender ratio and racial diversity of our society. This goal will require us to conduct more outreach activities to encourage youth to enter natural-resource professions. To recruit diversity into the work force, we must be certain to make all candidates comfortable in the search and interview process and provide checks and balances to assure no discrimination, blatant or inadvertent occurs.

## REFERENCE

Fox, K. A. 1993. A question of identity. Science 262:1090.

## COMMONLY USED TERMS

**Sex discrimination**. A behavior that treats a person differently because of his/her sex. That is to make employment decisions based primarily on an employees sex instead of his/her qualifications.

**Sexual harassment**. Unwelcome behavior of a sexual nature.

**Hostile work environment**. unwelcome sexual behavior that makes the work place offensive, hostile, or intimidating

**Third party sexual harassment**. Unwelcome sexual behavior that is not directed at an individual personally, but is part of the individuals work environment

## Chapter 5

# Landing a State Agency Job

LINDA ERICKSON-EASTWOOD AND FRED A. HARRIS

Working for a state agency can be a challenging and rewarding experience. Most state fish and wildlife agencies are responsible for the management and protection of aquatic life (primarily fishes) and ecosystems within their state. As a result, state employees work with a variety of aquatic habitats that support diverse communities.

### TYPES OF OPPORTUNITIES

State agencies usually have several types of positions. Positions range from part-time, to entry-level technicians, middle management supervisors, and administrators. Positions are commonly associated with research, management, or fish culture. However, many agencies now offer multi-disciplinary positions (planners, biometrician, aquatic education coordinators). Multi-disciplinary positions integrate knowledge in natural resources with knowledge in related fields (public information, mathematics, education, and computers). These positions are often not filled through the fisheries discipline. So, make sure to do some digging around in other disciplines (Ecological Services, Planning, Waters, etc.) to find these overlooked jobs.

To find out what types of jobs are available, keep in constant contact with the agency headquarters and/or staff within the discipline you wish to work. Many states have career bulletins or mailing lists. Also, check for job postings at local agency offices and state job service.

### EDUCATION

The level of education and training needed for each position varies among agencies. Position requirements can range from a two-year associate of science degree to a specialized doctorate degree. Most entry level fish culture and management positions require an associate's or a bachelor's degree. Most upper level management and research positions require a master's degree or combination of experience and education equivalent to a master's degree. Multi-disciplinary positions require a bachelor's in a related field combined with aquatic course work. Or, they may require a bachelor's or master's degree in fisheries with skills or additional courses in selected fields. A bachelor's degree will satisfy the minimum requirements for most jobs; however, applicants having a master's degree will likely be more competitive for these positions.

While earning your degree, you can increase your chances of being selected by incorporating these six items into your career plans:

1. Diversify your educational background. Combine fisheries and aquatic biology classes with those in human relations, sociology, planning, advanced mathematics, computer applications, media, business, and education (see Chapter 1).

2. Satisfy the current educational requirements for a Fisheries Scientist certification by the American Fisheries Society (AFS) (see Chapters 1 and 3). This certification guarantees that you get experiences viewed by employers as qualities of a well-rounded fishery's employee. It can also serve as a "tie breaker" during a tough hiring decision.

3. Become a skilled speaker. Staff frequently speak to a variety of groups (civic and angling clubs, legislatures, scientific societies). To be effective in your job, you must clearly express ideas and technical concepts in nontechnical terms. During the interview, a good speaker can quickly organize and convey their thoughts to the interview panel and that potential supervisor.

4. Develop writing skills. Good writers are held in high esteem. As in public speaking, you must express ideas and technical concepts to a diversity of audiences. One major benefit of a master's degree is the writing experience you get from completion of a thesis.

5. Obtain as much field experience as possible. State agencies look for applicants who have done a variety of duties and have been exposed to a variety of fisheries equipment. Gain experience by conducting a field project, by helping others on

their projects, by interning, and by volunteering.

6. Attend professional meetings to gain knowledge about your profession and to meet potential employers.

## OBTAINING INFORMATION

Start your search by contacting various state agency headquarters. Request information on staff size and ethnic/gender diversity, salaries, hiring procedures, organizational chart, advancement opportunities, job descriptions, and fringe benefits. A listing of state agencies, including mailing addresses and telephone numbers, can be found in the Conservation Directory, published annually by the National Wildlife Federation. Address your inquiry to their human resources discipline.

Once you have the above information, contact the potential supervisor directly to get detailed information on program goals, summaries of programs, related projects, and work hours and expectations for a specific job.

Participation in AFS provides avenues for obtaining contacts and information about state agencies and jobs. You will have opportunities to meet informally with many state agency staff and administrators at AFS meetings. These people facilitate future contacts. They also give you an "insiders" advantage to hearing about job openings and to knowing what skills a candidate needs to land the job. Your major professor or previous employer probably knows several key people. Ask him or her to introduce you to them.

## SELECTING AN AGENCY

Once you have obtained comparable information from several states, critically review them as to your career goals and job expectations. States differ in program focus, entry-level positions, career ladders, salaries, and promotion policies. Use the information to identify agencies that need your skills and abilities and provide job opportunities that satisfy your career interests.

Agencies have unique organizational cultures within which their employees function. While it is often difficult for prospective employees to become familiar with these values, they can be major determinants to job satisfaction following employment. Informal conversations with current employees can provide valuable insights into that agency's culture.

Your free time is as important as the job. So, don't forget to check out the area and community where you will be living. Often you will be moving to a different part of the country. Check state taxes, cost of living (insurance, home costs, groceries), schools, the community, geographic location, and job opportunities for your spouse by contacting the local chamber of commerce.

## HIRING PROCEDURES

Hiring procedures vary by state. Many states recruit new employees for lower-level positions and then promote personnel from these positions into higher levels. Others may require you to be a state resident before you can work for them. Still, others may not allow spouses or relatives to work for the same state agency. You should become familiar with the particular practices and requirements of those states with which you are seeking employment. Their human resources department will have this information. Then take special care to tailor your applications accordingly.

A current resume clearly identifying major academic course work, job related skills, and work experience is an integral part of any application (see Chapter 2). Many state agencies look for people who are "jacks of all trades." So, identify such skills as construction, electrical work, operation of machinery, or supervision on your resume. These skills could separate you from a host of other applicants. For natural resources jobs, don't follow the "industry rule" that a resume should be no longer than one page; instead include everything that you have ever done that might relate.

Many states "weed out" applicants by using a testing procedure. This process may mean scoring your resume, getting a list of take-home questions, or participating in a formal "test" situation. The best way to prepare for a test or interview is to review topics specific to the job. Find these topics by reviewing the job description and duties, by talking to staff, or by contacting the supervisor. Take this information and form some typical questions. Then practice answering the questions and study those areas you lack experience or education.

If asked, on an application or during an interview, give careful thought to the employment conditions and geographic locations for which you will accept employment. Often, failure to accept a position where you indicated availability can remove you from future employment lists. Too many restrictions can limit your opportunities for the scarce jobs available.

Once the interview is over, a potential supervisor will contact references. Choose these individuals carefully. It is in your best interest to select individuals that you have had recent contact with or know your abilities very well. They will do the best sell job. Don't forget to ask for their permission before using their names.

Remember, the current data privacy act allows references and agencies to only share the facts about your job and not judgements about your performance or abilities. Always contact your references and previous employers before an interview. Provide references with a current resume and the job description. Provide both references and previous employers with a written release saying that they may share performance information. This information will help them better prepare and give responses to those perspective employers' questions.

Remember, it is your responsibility to stay current on job opportunities and to notify agencies of any changes affecting your availability. Notification of address changes is important to remaining on mailing lists for job postings or tests.

Finally, you must be persistent. Don't expect to land that "premier" full-time job right away. Instead, look at ways to use part-time, temporary, and other full-time opportunities available to you. Take some risks. Move down or sideways on the career ladder for other experiences and possible avenues that will help you reach your lifetime goals.

Chapter 6

# U.S. Federal Employment

NEIL B. ARMANTROUT AND HANNIBAL BOLTON

The U.S. Federal Government is the largest single employer of fisheries and aquatic biologists in the U.S. The types of assignments are as varied as the profession, and are found throughout the U.S. and its Territories. Many Federal agencies maintain active foreign programs as well. The work may be in management, in research, or in regulatory affairs, and includes temporary and permanent positions. In general, the temporary and entry-level positions are primarily field positions, with the level of complexity increasing with experience.

A problem in finding employment with the Federal Government is the lack of a central hiring location. Each agency, and each office of an agency, often does its own hiring. Announcements of positions, particularly for temporary or entry-level positions, may only be advertised in a limited geographic area. Persons seeking employment should decide on the type of work and geographic location they prefer, then contact offices in that area on position availability. Since the number of people hired, particularly for temporary positions, fluctuates with changing budgets, several checks may be necessary. A central job opportunity board is available electronically, but may not carry temporary announcements or announcements only advertised locally.

## WHO EMPLOYS BIOLOGISTS?

Many Federal agencies employ fisheries and aquatic biologists. The type of work performed, number of biologists employed, and the location of employment varies by agency mission. Below is a brief description of some of the agencies that employ biologists.

*U.S. Forest Service*. The Forest Service manages land in nearly all states and U.S. Territories. The primary interest is in habitat management within an overall multiple resource use program. In addition, the Forest Service has a research arm and an active international support program.

*Bureau of Land Management*. The BLM is also a land management agency involved in multiple resource management. Most lands managed by BLM are west of the 100th parallel, in the Rocky Mountain and Pacific Coast states.

*U.S. Fish and Wildlife Service*. The FWS provides support to other federal agencies. In addition, it manages the nation's refuge system, conducts fish and wildlife research, is responsible for non-marine threatened and endangered species management, investigates violations of fish and game trade laws, and conducts an active international program. The FWS has offices throughout the U.S..

*National Biological Service*. The NBS was formed in 1993 by combining many of the research programs within the Department of Interior. It is active in research and surveying biological resources throughout the U.S..

*National Marine Fisheries Service*. The NMFS has the primary responsibility for management of marine fisheries resources, including threatened and endangered marine species. The NMFS is involved in research and resource management, and has an active international program. Offices are found in coastal areas.

*Environmental Protection Agency*. The EPA carries out basic research, and is active in environmental monitoring in many areas of the U.S..

*Department of Defense*. The Department maintains an active management program for resources on military reservations. The Army Corps of Engineers, which constructs and manages water projects, is a major employer of biologists within the Department of Defense.

*Bureau of Indian Affairs*. The BIA is involved in management of resources on lands owned by Native Americans, and on non-Native American lands where tribes retain treaty rights. Some biologists are hired directly by BIA, while others are employed by the tribes.

*Power Management Agencies*. The Bonneville Power Administration, Tennessee Valley Authority, and other power management organizations hire biologists to manage fisheries resources as part of larger power management programs.

*Bureau of Reclamation*. The Bureau, which is located mainly in the western U.S., manages fisheries as part of larger development projects.

Other agencies, such as Department of Energy (DOE), Aid for International Development (AID), and Natural Resource Conservation Service (NRCS) hire biologists for management, research, and development services.

## TYPES OF POSITIONS

Positions within the Federal service are defined and filled under Office of Personnel Management (OPM) regulations. Each skill category, such as fisheries biologist or aquatic biologist, is assigned to a series and grade. For fisheries and aquatic biologists, the series is GS-401, GS-482, and GS-486. The series describes the general type of position. The grade is based on the duties and responsibilities, with a grade of 01 being lowest. Most entry level positions for professionals are at a grade of five or seven. Requirements for any position are both educational and experiential.

The basic types of positions include technical, professional and managerial. Professional and managerial are career positions; promotion depends on successfully performing at the next lower level. Technicians perform work supporting the operation of scientific programs. They may be temporary or permanent.

Positions can be permanent full time, permanent part time, for a specified length of time (term), or temporary. Persons first entering Federal service in a professional series are probational for the first year. After three years of successful job performance, employees may be granted career status.

## COMPETING FOR POSITIONS

Most permanent jobs are filled competitively. Positions are identified by an individual agency office. The duties to be performed and the job requirements are written into a Position Description. This Position Description is used to determine the type of appointment and grade level. Positions are then advertised in a Vacancy Announcement. This Announcement briefly describes the duties, position requirements, and application procedures. Each Announcement has a number and a contact office. Since many Announcements are open only for a short period, it is important to keep checking posted Vacancy Announcements. Advertisement is usually for one to three weeks (but may be longer or shorter), and can be local, regional, or national, depending on the grade and complexity of the work. While most positions are opened to all qualified applicants, some may be restricted to a particular group of candidates. This is usually due to hiring restrictions, and is stated in the Vacancy Announcement.

Federal Job Information Centers provide regional job information, job announcements, qualification requirements, application forms, and addresses of all Federal agencies. For entry-level and temporary positions, the positions are often advertised only on a limited geographic basis, so it may be necessary to contact more than one Federal Job Information Center. Agencies can be contacted directly for information on positions, particularly those in a local area. For some temporary positions, hiring is now done through State Employment Offices. Information on these positions is available from the hiring agency, State Employment Office, or Federal Job Information Center.

For many years, Federal agencies utilized the SF-171 as the application form. It is no longer required, but may be used as on option. Applicants may submit Option Form 612, or a standard resume. While there is no set form for applying for a Federal position, there are a number of items that must be included in your application. You can contact any Federal Job Information Center for a list of the required information, or download the information from the electronic bulletin board. Be certain to read the Vacancy Announcement application requirements. You should send in only the information requested. In addition to the application form, some require additional information, such as a transcript of classes, a performance rating from previous employment, or documentation of particular skills or experience.

Your application is a most important document - it is your best opportunity to show why you are the best qualified for a position. As with any application, it should be accurate, with correct spelling and grammar, typed if possible (computer programs are available for the SF-171, and soon to be available for Option Form 612). Appearance is important. Include all pertinent information, including details. For example, don't just say you did inventories: provide details, such as the type of inventories, equipment used, use of computer, reports and recommendations prepared, etc.. Certain types of experience, such as supervisory, analysis and evaluation, public interaction, use of specialized equipment, etc., can be important in improving your rating. The initial evaluations of your application are done by people who are not biologists, who are using general guidelines, and who will be looking for statements that show you meet the qualifications for the position. Volunteer work related to a position is given the same credit as paid experience.

Persons applying for openings compete with all other applicants. The better you are at showing your capabilities, the better your chances for being hired. Most announcements include a section called KSA, Knowledge, Skills and Abilities. Since they relate directly to the duties of the advertised position, these items will receive the greatest emphasis during the review of applications. If you have questions or wish more information, you may contact the office that is doing the hiring. Showing an interest in a position can be a help.

Applications are used to rate the applicants. They are listed as best qualified, qualified, or not qualified. For some professional positions, particularly at higher grades, a panel may be formed of professionals that will rate all qualified applicants. A list of eligible candidates is used to prepare an eligibility roster, which is then sent to the person making the final selection.

Requirements for employment vary according to the position, but there are two skills that are widely required: A basic knowledge of fishery science, and the ability to communicate well orally and in writing. In addition, there are several types of skills widely requested.

- Broad knowledge of multiple resource issues, and the ability to view fishery issues as one resource in a mix of resources.
- Ability to gather and analyze information, and to do evaluations and impact analysis.
- Knowledge of Federal programs such as the Threatened and Endangered Species Act and water quality issues.
- Concepts of biodiversity and ecosystem function and management.
- Knowledge of computers and data-base management.

Most Federal agencies rely on temporary employees to carry

out field assignments. Temporary positions provide good training and experience. They are often a good avenue for permanent positions, since persons performing well as temporaries are often hired for permanent positions. Other programs, such as the Youth Conservation Corp (YCC) and volunteering, also provide experience and help in later obtaining a permanent position. The YCC program is for students 15 to 19 years of age, and is administered by a government agency. For volunteering, there are established programs, such as the Student Conservation Association, but individuals can contact the Federal office of their choice and volunteer directly.

## NON-COMPETITIVE EMPLOYMENT

The Co-op Education program allows students to be hired non-competitively after receiving a degree, although they must initially compete to become part of the Co-op Education program. The purpose of the Co-op Education program is to provide experience to qualified students that will make them eligible for permanent employment after they receive their degrees. To be eligible students must be enrolled in a participating college. Co-op Education openings are advertised. Students wishing to participate submit an application. After being selected, the student attends school part of the time and works for the agency part of the time. After graduation, the student may be hired to a permanent position.

Former Peace Corps or Vista volunteers may be appointed to permanent Federal positions within a year after successfully completing volunteer service. This may be extended to two years if the former volunteer attended school. Nondisabled veterans with no more than 16 years of education who served on active duty after the Vietnam War, and who meet other eligibility requirements, may be given a Veterans Readjustment Appointment (VRA), which is a 2-year appointment during which the veteran continues education and training. After the 2-year VRA, the veteran may be converted to a permanent appointment. Qualifying disabled veterans may also be eligible for temporary appointments, which may be converted to permanent positions.

## FURTHER INFORMATION

There is a Federal Job Information Center in each state. They are usually in the largest city in the state. Check the listings under Federal Government in the phone book for addresses and phone numbers. You may request the booklet *Applying for a Federal Job*, which briefly discusses the procedures for applying for Federal employment, and summarizes general information required in your application.

The Federal Job Opportunities Board is an electronic bulletin board system that makes job information available for viewing as well as downloading to personal computers. It operates 24 hours a day, seven days a week. The modem access number is 912-757-3100.

Career America Connection Telephone System includes a College Hotline and General Jobline, using a customer-directed search routine with pre-recorded messages. The access number is 912-757-3000.

Many organizations, including the American Fisheries Society, provide job listings. Two private companies, Federal Jobs Digest (1-800-824-5000, ext. 1116) and Federal Research Service, Inc. (703-281-0200) have on-line information compiled from Federal listings.

Chapter 7

# Fisheries Employment Outside the U.S. and Canada

GENE R. HUNTSMAN AND RICHARD A. NEAL

## IS A FOREIGN ASSIGNMENT FOR YOU?

Choosing a foreign assignment over a domestic one can be a greater than-ordinary decision as it affects your career in fishery science. A choice of a foreign position has the potential to radically alter your future. When the job, your personality, and collateral factors like family and romantic ties all mesh together well, foreign assignments can be immensely rewarding and provide you with a ready pathway towards international prominence in fishery science. On the other hand, foreign assignments have greater potential to present unexpected problems than do jobs in your homeland wherein you are supported by your native culture. To provide guidance about the suitability of a foreign position for you, we offer the following discussion of the advantages and disadvantages of assignments outside of the U.S. and Canada.

### Advantages

The advantages of a position outside the U.S. or Canada can be substantial and range in nature from philosophical to financial. Working abroad almost inevitably engenders contact with animals, ecological systems, and often, problems that are different from those in your home country. Thus, primary advantages of foreign employment are intellectual stimulation and the broadening of your knowledge about fisheries. Different insights and perspectives obtained while working abroad can make you a better biologist for the remainder of your career, even if that remainder is spent domestically. If your ambition is achievement of an international reputation in fisheries science there is hardly a better vehicle than assignments abroad. They allow you to meet and interact with peers from many nations. Often because of the small size of, greater connectedness between, and a more internationaistic philosophies of, countries abroad, your opportunity for international interchange is, in most cases, proportionately much greater if you hold an overseas position. Foreign assignments in many cases make a biologist more competitive for domestic employment. Experience in foreign countries may provide the worker with a mystique of worldliness attractive to employers.

A more directly venal reason for accepting a foreign assignment is that for one or a combination of reasons the remuneration can, in some cases, be better than for available domestic positions. The reasons can include:

- the possibility of higher pay than for equivalent domestic work;
- substantial perquisites;
- lower cost of living in some countries;
- tax advantages pertaining to money earned out of the home country.

Often the basic pay is greater with an international agency, a foreign university, or private firm abroad than is pay for equivalent domestic work. Employees of the U.S. government, when holding assignments in international agencies to which the U.S. is party, receive remuneration from either the international agency or the U.S. government (whichever pays higher) which includes not only net pay, but also post allowances and differential, living quarters allowance, and a dependents' educational allowance.

The fringe benefits and perquisites attached to foreign positions are often substantial. A common fringe benefit is annual paid travel for the employee and, sometimes, dependents to and from the home country. Vacation time is usually ample, and such amenities as housing may be furnished. United States government employees temporarily attached to international agencies retain the health and life insurance benefits of their employing agencies, as well as reemployment rights, retirement credit and other benefits.

A third benefit accruing to workers abroad may be lower living costs. Especially in developing countries, cost of food, housing, and domestic servants may be extraordinarily lower. Families living in developing countries may achieve a lifestyle available only to the very wealthy in the U.S. and Canada. On the other hand, living expenses in some countries are higher than the U.S. and Canada. While salaries abroad may be adjusted upwards to match increased costs, the net surplus of income over costs may be the same as at home. Again each employment situation differs. A prudent prospective employee will make a careful examination of costs, they might encounter when working abroad.

Canadian and U.S. citizens employed abroad by foreign governments, universities, and private businesses (but, in the case of U.S. citizens, not by the U.S. government) may be eligible for substantial tax benefits if the condition of non-residency is established in the home country. In 1993, U.S. citizens could exempt up to $70,000 in income from U.S. tax liability if they were out of the United States for an entire tax year, or for 330 full days in any consecutive 12-month period. Non-residency may be harder to acquire for Canadians. Eligibility hinges on possession of real and financial properties in Canada, family ties, and other issues. Because tax rules are volatile, anyone considering foreign employment should check policies at the time of employment. Income acquired abroad may be subject to taxes in the country of residency.

A fifth financial (and often intellectual) benefit of working abroad is the opportunity to travel in and near the country of employment for a fraction of the cost that would be incurred were Canada or the U.S. the home base. Additionally, flights to and from the country where employed are often available with alternative routes and offer free or inexpensive stop-over privileges. Thus, global travel subsidized by the employer may be possible.

**Disadvantages**

The old principle of "no free lunch" applies to foreign employment as well as ecology. The benefits of working abroad can come hand-in-hand with adversities that, depending on the employee and her/his subjective evaluation, range from amusing challenges to insurmountable and unacceptable difficulties. Again prior to accepting employment, a candidate ought to investigate conditions in the country of future residency to the greatest extent possible.

While it is impossible to predict all of the impediments likely to be encountered in an overseas assignment, we will review a few of the most prominent. Perhaps the most significant contributors to culture shock are psychological, philosophical, and otherwise abstract aspects of life in a foreign country. For professionals working abroad a common source of disillusionment is encounters with bureaucracies that seem even more inscrutable, multilayered and counter-productive than that in the home country. While the government structure may fit well with the expectations and attitudes of the permanent inhabitants, its functioning may appear arcane, and perhaps, non-existent to the North American mind. Given that fisheries professionals often are intimately involved with the local government, patience, diplomacy, flexibility, and adaptability are good qualities for such professionals. Again some overseas governments are extremely effective and efficient. A valuable cliché' is "look before you leap." A wrong choice could result in major frustration on your part.

In addition to government structure, overall cultural attitudes in the host country can greatly affect your success and happiness while in your foreign assignment. Perhaps the most important attitudes are those about:

- work;
- women in the workplace;
- alcohol and drugs;
- sex and gender relationships.

While zeal for labor in most countries equals or exceeds that in Canada or the U.S., industriousness, as a cultural attribute, is not universal. Attitudes about work intensity, work hours, and absenteeism, while perfectly acceptable to the local people, may not be what the visiting North American expects. Completing projects and otherwise meeting deadlines may not be as important to indigenous cohorts and assistants as to the visiting investigator. Again frustration and disappointment may be avoided by a realistic appraisal of conditions before accepting a position or scheduling completions.

The position of women in the professional workforce in North America has advanced markedly in the last decade, but such advances have not occurred world-wide. There are cultures abroad wherein women will be at a distinct disadvantage to men when working as a fisheries professional. Even when women working abroad find relative equality and respect among native male co-workers, many of whom may be western-trained, these same women may not enjoy freedom of movement and expression among the general populace.

Attitudes about use of alcohol vary widely among cultures, and could create pitfalls for a foreign investigator. Where heavy use of alcohol is common, the condition of workers may occasionally suffer and absenteeism may be greater than expected. Conversely in some, principally Muslim, countries, use of alcohol is forbidden even to foreigners, and visiting investigators risk social ostracism or legal action if they indulge in even modest use of alcoholic beverages. Recreational use of some psychotropic drugs is tolerated, if not legal, in some parts of the United States. That same usage, or even possession of minor amounts of drugs, in other countries could result in long imprisonment under primitive conditions, or even death. Thus visitors must be especially heedful of local laws about drugs. Similarly not all cultures in the world share the relatively liberal Euro-American attitude about sex and gender relationships. Adultery still merits death, perhaps by stoning, in some cultures. On the other hand, some cultures are much more relaxed about sexual relationships. The reader is left to ponder the problems and opportunities this circumstance presents.

A major contributor to culture shock can be the necessity to function in a second language. The need for language skills is usually clearly defined in employment announcements, and most foreign workers possess sufficient knowledge of the required language to function professionally, often among English speaking coworkers. But often the meager language skills of the visiting professional do not allow her/him to function fluently on the street, in social gatherings, or the marketplace. Lack of fluency can establish a sense of isolation and alienation between the visiting researcher and the local inhabitants. On the other hand, there is no finer training in a foreign language than that provided by necessity.

A second major source of adversity peculiar to some foreign positions is increased risk to physical health and welfare. Perhaps particularly in some tropical countries, there are a host of diseases not found in North America and for which special preparations must be made, or which require some out-of-the-ordinary diligence to avoid. Chlorea, malaria, plague, yellow fever, etc. still exist in some parts of the world. By and large, the worst illnesses are prevented by inoculations required, or recommended, before initiation of travel. Common sense will prevent others. The high frequency of HIV infections in some African nations, for instance, indicates that extreme care must be attached to sexual activity and transfusions of blood.

Air pollution levels in the larger cities of the developing world may be of concern, especially to persons with respiratory problems. Potable water may not be readily available, and provision of such water could become a major activity. Standards of sanitation in storage and sale of fresh meats and produce and disposal of sewage and waste, are very different in many countries. Conditions are often labelled unhygienic by North Americans, but while prudence in hygiene is always good, most western digestive tracts are capable of dealing with more microbes than the proprietors of those tracts would believe.

In all remote areas, especially of developing nations, medical care may not have yet achieved the standards of North American practice. Modern medical facilities probably are available within a few hours flight of most airports, however.

As in the U.S., crime rates can be much higher in some areas than others. Unlike the North American norm, activity designed to avoid theft and muggings can be a major part of a day's activities in some areas. Elaborate physical barriers (walls topped by broken glass, etc) to prevent theft are an attribute of most homes in crime prone regions. Foreigners, especially those clearly identifiable as tourists, because of their suspected affluence, are the preferred target of muggers in some regions. Political instability and engendered civil strife, terrorism, and even war are, unfortunately, characteristic of some regions. The risks posed by violent political unrest are obvious. Again westerners (especially government workers) are sometimes preferentially selected for kidnaping, injury, or murder.

Less likely to injure a person's body, but more likely to cause injury to research and to personal comfort are deficiencies in a category we call logistics. Electricity is, of course, absent from many rural regions of the world. Without electricity, air conditioning, an essential for many twentieth century westerners in the tropics, cannot be had. Even where electricity is provided as line-current its availability may be irregular. Murphy's law works world-wide, and current will choose the most impropitious times to disappear. While instruments for research, computers, outboard motors and vehicles are usually available, these necessities may suffer inordinate (by North American standards) time out of service because spare parts or skilled labor for repairs are hard or slow to acquire. Communication facilities, especially telephone and mail service, while much improved in the last decade, are neither reliable or fast in all areas. Deficiencies in communication can impede research and contribute to a sense of isolation. Finally, and not necessarily trivially, the absence or scarcity of personal amenities can assume large proportions in measuring the quality of life overseas.

An important factor in the response of any worker to conditions overseas is her/his family arrangement. In some positions and areas the worker who has a family with her/him is somewhat insulated from life in an alien culture and perhaps happier than a single person while responding to the difficulties of a new lifestyle. Conversely, some assignments entail environmental demands of such rigor that presence of a family, especially with small children, would engender debilitating concern about their health and safety. Provision of education for children may be difficult. Unmarried workers abroad, because of cultural obstacles, sometimes find such difficulty in forming romantic relationships in the host country that they must leave an otherwise satisfying position to establish a family.

Of course, not all, or any, of the potential difficulties we have listed will occur at any particular position. We provide the listing to show that resources which are taken for granted in North America may not be available elsewhere. The deficiencies do not have to detract from the enjoyment of a foreign assignment. If the changes in lifestyle and facilities are viewed as challenges, their overcoming will add richness to your life and to memories of the experience.

Preparing to choose an overseas assignment requires substantial research. Interview other North Americans working in the area. Investigate medical and political conditions. Having accepted a foreign position try to be prepared and alert, but not surprised.

For political, economic and practical reasons almost every government would prefer to hire its own citizens rather that hire foreign nationals. Further, for similar reasons, international organizations and private groups operating in other countries will also hire local personnel whenever available. The person interested in working outside the U.S. and Canada must, therefore, have something special to offer in the way of training, experience or capabilities. Journeymen scientists, technicians, and teachers usually will not have opportunities for employment overseas as locals with adequate skills and training are available. Some exceptions to this practice are the hiring of unskilled or semi-skilled persons for Peace Corps, church, or other voluntary organizations that are active with low-technology assistance in impoverished communities. The tasks of these volunteers may be general community assistance activities including some fishing or aquaculture support.

An experienced fishery biologist, economist, sociologist or administrator with at least ten years of experience and a clear area of specialization may find opportunities for foreign employment, particularly if the area of specialization represents new applications of technology. Also, knowledge of the language of the country is usually important as is experience working in foreign countries. In developing countries, where most employment opportunities exist, it is important that the level of skills of the applicant match the level of application required. Computer facilities or laboratory equipment available may be less sophisticated than those required for many experts to contribute.

## SPECIAL QUALIFICATIONS DESIRED

For any foreign assignment several special personal qualifications are important. Among these are: flexibility - a capacity to promote ideas vigorously without taking rigid stances; adaptability - the ability to adjust easily to drastic changes in working environment, information bases, cultural values and expectations; good health - additional stress, air pollution, gastro-intestinal disturbances and illness must be handled easily; and compatibility - working closely with people who view things differently is a part of most foreign employment.

Although some research positions are available overseas in fishery science, openings are more likely to be related directly to fisheries management, to collection of data related to management, to aquacultural production, and to teaching fishery science. For aquaculture new technology exists in relation to genetics, reproduction, nutrition and diseases. Leaders in these specialty areas may be in considerable demand around the world.

Statisticians, computer programmers, stock assessment experts, and modelers are widely in demand, particularly in developing countries. Personal computers are now widely available, even in developing countries, but training in full use of these computer capabilities is needed. Application of available software, programming to meet specific needs, and interpretation of the results of software applications are areas where assistance will be useful. Statisticians who can provide leadership in experimental design and development of fishery sampling protocols may have opportunities for employment abroad.

As trade in fishery products increases and hygienic standards receive more public attention, industrial assistance in the handling, packaging, and preservation of fishery products is being sought by commercial firms as well as some governmental and university groups. The development and marketing of new specialty fishery products is a related area attracting interest.

Fishing expertise, particularly expertise in the use of new gear is needed by some countries. Examples of fishing expertise needed by countries developing new fisheries are longlining, use of acoustic equipment, and skills in pelagic and midwater trawling although for the larger, expensive gear the supplier often provides training in its use.

Fishery managers need supporting personnel who can help them justify and promote appropriate management measures to ensure sustainable, economically sound use of resources. In all countries businessmen and politicians tend to favor a short-term view of resource exploitation vs. longer-term sustainable use. Specialists needed to assist management include stock assessment biologists, fishery economists, and increasingly, sociologists. The special role of economists in working to limit harvesting capacity has not yet been applied in most countries and particularly in developing countries, social concerns, such as employment, often override the efforts to improve economic efficiency of the industry.

Environmental concerns are growing around the world; however, fisheries expertise is not often brought into efforts to solve environmental problems. Skills in organizing and addressing large environmental issues on a political, social, economic and biological basis are in demand.

Some fisheries administrators with a depth of experience in organizing and implementing complex programs are hired for foreign positions.

Experience in foreign countries is an asset in these assignments as is successful administration of fisheries research or management organizations.

## POTENTIAL EMPLOYERS

### Foreign Governments

Foreign governments may require special expertise or exceptionally capable individuals for selected expert or advisory roles. Poorer countries are less likely to have nationals with all the skills needed, particularly those relating to new technologies. These positions are often temporary or short-term and are likely to include technology transfer and training activities. Positions are usually within the fisheries department of the country involved. The poorer countries typically have low government pay scales making these positions less attractive. Some countries utilize international organizations or consulting firms through contractual or other arrangements to hire foreigners at international rates thereby avoiding national limitations on pay scales.

### International Organizations

International organizations established to promote cooperation, development or rational resource use are the principal sources of foreign opportunities for employment. The Food and Agricultural Organization (FAO) of the United Nations for example, has an active fisheries program of broad scope through which development assistance is provided. Other groups may stress research (e.g., the International Center for Aquatic Resource Management, ICLARM), environmental concerns (e.g., United Nations Environmental Program, UNEP), management of shared resources (e.g., the South Pacific Commission, SPC) or development loans (e.g., Asian Development Bank). Experts in a wide variety of fisheries-related fields may find interesting opportunities with these groups. Pay scales and benefits are designed to be attractive to experienced fisheries experts from developed countries.

### Private Voluntary Organizations

These organizations serve a wide variety of purposes ranging from community development to resource protection and their organization is similarly diverse. Often, these groups promote development, and some include fisheries or aquaculture in their work. Both those with considerable experience and the inexperienced may find opportunities with these groups. Typically pay is low for assignments in developing countries, but benefits provided, such as food, room, and vehicle, may compensate for low salaries.

Work with a private voluntary organization is often a route to foreign work with other organizations as it provides experience in a foreign setting, and introduction to foreign cultures.

**Consulting Firms**

Some consulting firms specialize in providing expertise and special assistance to foreign governments, U.S. government agencies and international organizations for both the short- and long-term. Those interested in hiring U.S. citizens have offices in the U.S.

Positions typically require substantial expertise, and duties may range from evaluations or reviews lasting two or three weeks to multiyear assignments. Project planning exercises are also often conducted by teams hired by these companies. Consulting firms also conduct long-term development projects for agencies such as the U.S. Agency for International Development, the World Bank, and the Canadian International Development Agency. Foreign governments may solicit specialized expertise through consulting firms. Some consulting firms work with one or more U.S. universities to provide needed expertise from their faculties and staffs. Personnel to train extension workers, fishing vessel observers, aquaculturists and various other professionals are in demand in some instances.

**Private Companies**

International commercial firms are active in foreign countries and may have openings for individuals with biological, fisheries or aquacultural experience. Joint fishing ventures may require expertise with gear, vessel operation and maintenance, or handling and processing of products. For example, the shrimp culture industry in Latin America had ties with U.S.. companies in its early days and hired hatchery, feed and disease experts from the U.S. The increasing international market for quality fish products has increased need for product handling experts. Skilled fishermen may be required to assist with purse seining, long-lining, or other specialized fishing operations.

Again, foreign employees need to have special expertise not readily available in the country concerned.

**Universities**

Many universities have an associated "institute" designed to be a business arm of the university and to handle contracts, commercial activities, and funding from outside sources for research. These institutes often hold development contracts in foreign countries, particularly if the contract includes research, training, or education and matches the expertise of faculty members in some special field such as fisheries or aquaculture. The institutes may also join forces with consulting firms as noted earlier to help supply the expertise needed to fulfill a particular development contract. The institutes tend to look for expertise first among their faculty and students, but may also hire non-university experts.

Universities (Auburn is an excellent example) may have strong ties with other countries or with foreign universities with arrangements to exchange faculty, educate graduate students, or provide long-term assistance of an educational or research nature. University teaching in the fisheries area is a frequent area of employment (e.g., The University of Oman has hired fishery scientists from the U.S.).

**Government Agencies**

The U.S. Peace Corps has placed large numbers of volunteers with some basic aquacultural training in developing countries. These volunteers tend to work in community development to introduce the concepts of aquaculture. The U.S. Agency for International Development and the Canadian International Development Agency fund some fishery and aquacultural projects that are normally staffed through contracts with consulting firms or with universities. The U.S. Department of State has Fishery Attache's in several overseas posts who serve as regional contacts for international negotiations on fisheries affairs and legal matters for the U.S. Government.

## SOURCES OF INFORMATION

Information about foreign positions is diffuse and locating it may require detective work on your part. A good source is the AFS jobs list published in *Fisheries*. The *Fisheries* listing is usually timely but not always exhaustive. A similar list is provided in the newsletter of the Aquatic Ecology Section of the Ecological Society of America. Classified advertisements in *Science* often list positions at foreign universities. Word of mouth can provide rapid notice, sometimes before formal annoucement, of jobs overseas but such notice usually requires a substantial network of well placed associates. Younger fishery workers may not have access to such a network.

Direct contact with employing agencies is always worth trying. Some of the most important employers are:

International Commission for Conservation of Atlantic Tunas
Principe de Vergara 17
E 28001 Madrid, Spain;

International Center for Living Aquatic Resource Management
MC PO Box 1501 PH:28180466, Makati, Metro Manila
Phillipines;

International Council for the Exploration of the Sea Palaegade
204 PH: 1114300
DK 1261 Copenhagen K, Denmark.

Domestic agencies often have international operations. Contacts should be made with:

Office of International Affairs
National Marine Fisheries Service, NOAA
U.S. Dept of Commerce
1315 East-West Highway
Silver Spring, Maryland 20910-3225;

International Development Research Centre
P.O. Box 8500
Ottawa, Ontario K160H9;

Canadian International Development Agency
(work often done by private firms under contract)
200 Promenade du Portage
Hull, Quebec K1A064;

U.S. Agency for International Development;

U.S. Department of State;

Many environmental Organizations sponsor work abroad. A leader is the:

World Wildlife Fund
1250 24th Street, N.W.
Washington, D.C. USA
Phone: 202-789-3000

Written works of use in seeking overseas employment include:

Kocher, E. 1993. International Jobs. Addison-Wesley, Publishing Company., New York, New York, 394 pp.

UNESCO. 1983. International Directory of Marine Scientists. United Nations Educational Scientific and Cultural Organization. Paris and The Food and Agricultural Organization of the United Nations, Rome. 173 pp.

Koek, K.E., editor. 1989. Encyclopedia of assocations; International organizations. Gale Research Inc. Detroit, Michigan, 1433 pp.

# Chapter 8

# Fisheries Employment in Canada

BRUCE A. BARTON

Canada is a vast country and rich in fisheries resources. However, the population and the economic base are low, and thus, the job market for fisheries biologists in Canada is more restricted than in the U.S.. Nevertheless, the diversity of aquatic resources and their associated problems throughout the country provide a wide array of potentially challenging opportunities. Traditionally, fisheries graduates have sought career jobs with federal and provincial government agencies. But in this time of shrinking government budgets and the trend toward "privatization", private consulting companies are becoming more important for providing career positions. Typical entry-level fisheries jobs with government agencies are often field positions in areas of resource inventory, habitat protection and development, fish culture, and law enforcement.

All fisheries management in Canada is carried out pursuant to the federal Fisheries Act, which is designed to protect and regulate all fishery resources in the country. The Fisheries Act legally empowers the federal government with the ultimate authority for all fisheries waters. Management of freshwater fisheries, however, is mainly the responsibility of the provincial agencies. Regulatory aspects of provincial programs must therefore be approved by the federal government to ensure compliance with the Fisheries Act. Management of marine resources is the sole domain of the federal government. Anadromous fisheries, notably salmonids on the east and west coasts, are managed jointly with the provincial agency being responsible for the freshwater component whereas the federal government assumes responsibility for the marine fishery. Thus, a high level of cooperation between the two agencies is required for effective management. Similarly, management of fisheries in the Great Lakes is carried out cooperatively by the Ontario provincial government and respective U.S. state agencies. Coordination of joint programs is the responsibility of the Great Lakes Fishery Commission.

Opportunities for employment in all aspects of fisheries (e.g., management, culture, environmental protection, and research) can occur at both federal and provincial levels. Unfortunately, entry-level positions in fisheries at either government level are increasingly difficult to find. Those who are better prepared to enter the job market will have a competitive advantage.

## THE PROVINCIAL AGENCIES

As provincial agencies are responsible for managing Canada's inland fisheries, they conduct resource inventories and research, operate hatcheries, develop and protect habitat, establish and enforce regulations, and provide public information programs for the variety of freshwater fish resources throughout the country. Thus provincial agencies are often the first choice for employment by fisheries graduates. The provincial government branches responsible for employing fisheries personnel in Canada are listed in appendix 8.1.

Job seekers should be aware, however, that many of the agencies, particularly those in larger or more heavily populated provinces, operate administratively from regional offices. In these cases, job hiring may be done at the regional level while "head office" provides mainly policy and program direction. Thus, the new job hunter is well-advised to visit with regional or district biologists and file job applications with the regional agency offices in the province of choice. Moreover, job applicants should not confine their search to only the fisheries management components of government. Opportunities for fisheries biologists may also exist with other government departments or divisions such as those responsible for parks, forests, lands, environment, water resources, and agriculture.

Provincial fish and wildlife officers are at the front line of communication between the agency and the public and deal with practical fisheries-related problems on a daily basis. For this reason, becoming a field officer may be an attractive choice for those wanting a career with a "hands-on" approach away from a desk. However, like any position with legal authority, an element of personal risk is assumed and special enforcement training is required beyond receiving a degree or technical diploma.

## THE FEDERAL GOVERNMENT

The major employers of fisheries biologists in the Canadian

federal government are the Department of Fisheries and Oceans (DFO) and the Department of the Environment (DOE). Both departments have regional administrative centers and research laboratories throughout Canada; regional centers and major research facilities of DFO are listed in appendices 8.2 and 8.3, respectively. In addition to the research establishments, other special programs exist that have specific mandates (e.g., the Salmonid Enhancement Program located in Vancouver, BC).

Habitat management and protection represents a source of employment opportunity for graduates interested in fisheries habitat problems. Much of the work of fish habitat biologists within DFO or DOE involves standards development and reviewing projects for compliance with environmental legislation. Maritime fisheries enforcement is the jurisdiction of the DFO and a potential career option. Like their provincial counterparts, federal fisheries officers must undergo special training beyond a traditional diploma or degree. Federal government departments that employ fisheries biologists on a smaller scale include Heritage Canada, which includes Canada's national parks.

Currently, the Canadian federal government is undergoing a major reorganization and redistribution of fisheries and habitat responsibilities. At the time of this chapter's preparation, best information indicates that all aspects of marine fisheries, including enforcement and habitat protection, will be the domain of DFO. Responsibility for freshwater fishery resources, including habitat, will reside with DOE or the provincial governments.

A fundamental step in getting any federal government job is completing the appropriate civil service application form (Form No. PSC 3181) and submitting it to a regional Public Service Commission of Canada (PSC) office in response to a posting or opening listed. The job seeker should check with the PSC office of choice (appendix 8.4) as hiring policies are made regionally and depend on the inventory of jobs available. Vacancies for federal positions with DFO or DOE should be posted with regional or district PSC offices and usually can be scanned either by computer or in hard copy at their major centers. Although recommended, this approach may not be particularly fruitful as federal vacancies are required to be advertised in-house first and may be filled by temporary or contract employees.

## THE PRIVATE SECTOR

Recent graduates and those looking for possible career changes should not overlook the potential opportunities that exist in the private sector. Government agencies are increasing their reliance on private consulting firms to carry out much of their work, particularly information gathering. Consulting work varies with the needs of the contractor. Duties may include desk-top literature reviews, toxicology studies in a lab, or conducting fish stock assessments in often very remote locations. Working for a private consultant can provide a diverse range of valuable practical experience in a relatively short period of time that the new graduate may not otherwise obtain.

Although consultant directories exist, an easy approach may be to simply use the Yellow Pages to establish a list of potential employers in the area. Be cautioned, however, that the majority of environmental or resource consultants listed in directories may focus on environmental engineering and not offer fisheries services. A good way to find out about respectable private consulting companies with fisheries expertise is to ask local government biologists as they have often been the contractor for such services in the past.

Large private organizations are also possible sources for career-oriented positions. Public utility corporations, such as Ontario Hydro or BC Hydro, hire fisheries biologists. Industries that depend on natural resources, such as the petrochemical or pulp and paper industries, also may hire fisheries staff if they are large enough to warrant it. The type of work within these organizations may be in-house research studies or development-proposal reviews aimed at regulatory compliance or licensing. Small companies usually contract this kind of work to the private consultants.

Career opportunities for fish biologists may be found in aquaculture. Private fish aquaculture is practiced mostly on the east and west coasts of Canada and in southern Ontario, with the emphasis being on salmonids for the food market. After a down-turn a few years ago, the Canadian aquaculture industry has stabilized and will likely expand in the future as commercial fish stocks continue to decline.

A fisheries instructor position at a technical or community

college is a possible career avenue for graduates at the master's level. However, new openings are infrequent and competition for these positions will likely include doctorate level applicants and those with considerable experience. Similarly, doctoral graduates will find new teaching positions at universities scarce and highly competitive.

## JOB REQUIREMENTS AND GETTING STARTED

Specific requirements for any fisheries biologist position are likely to be as varied as the job it serves. Educational requirements range from a technical diploma or bachelor's degree for a fish technician, to a master's or doctorate degree for a senior research-oriented position. No formal requirements are stated but a degree in biological sciences with an emphasis on fisheries or aquatic sciences is highly recommended. Unlike many U.S. counterparts, universities in Canada do not have separate departments for fisheries and wildlife sciences. Appropriate fisheries-related undergraduate curricula are available nation-wide with at least one institution in every province offering some level of fisheries-related instruction. Canadian employers have not expressed much interest in requiring American Fisheries Society (AFS) professional certification, but meeting the minimum academic standard for certification will nevertheless provide a good foundation for job market entry or graduate studies in fisheries. Certainly, certification with AFS or another professional society can be a positive addition to one's curriculum vitae. Membership in a legally recognized professional biologist society in those provinces that have them would also be an asset (e.g., British Columbia, Alberta).

Although it may not be stated officially, a master's degree is now almost a must for any entry-level career position in fisheries management with a government agency. While this may seem unfair to recent bachelor graduates eager to start their careers, the fact is that managing fisheries today is much more complex than it was a few decades ago. The additional experience obtained conducting independent research beyond the undergraduate level is invaluable for learning skills in data analysis and presentation, field or laboratory techniques, and especially, critical thinking. Also, attending professional meetings as a graduate student and presenting one's own research results are excellent ways to make new employer contacts.

The adage, "there's no substitute for experience", certainly holds true for the fisheries profession. Perhaps the best way to get experience and establish contacts with future employers is through a summer job. Summer employment in fisheries may involve performing inventories, conducting creel surveys or restoring habitat, and can lead to longer-term contract positions after graduation. When looking for a summer job, start as early as the previous year as seasonal positions in fisheries are usually scarce and very competitive. Technical and personal skills are useful attributes to include in your résumé along with your academic and employment background. That first successful application for a seasonal job may depend as much on familiarity with boats and motors or computer spreadsheets as on formal university instruction.

The employment outlook in Canada for entry-level fisheries biologists presently is not very good. All government agencies, particularly the federal government, are "down-sizing" in order to reduce their employee numbers and budget deficits. Moreover, employees declared "surplus" in other federal positions as a result of staff reductions will have priority under special provisions for filling new job openings. Natural resources management sectors of government appear to be most affected by these cost-cutting measures. The reality of the current situation is that all entry-level positions are now likely to be temporary and seasonal. Thus, the chances of the new graduate stepping directly into a permanent position in fisheries are remote. Similarly in the private sector, jobs in fisheries are often temporary reflecting the nature of contract awards to consulting firms, many of which originate from the government. The apparent lack of career employment opportunities in Canada may be initially discouraging. Accepting temporary positions, however, is often the best and only way to accumulate that much-needed experience and eventually land a permanent career position as a fisheries biologist.

## ACKNOWLEDGMENTS

Information and/or review comments were provided by: William Franzin (MB), Duane Radford (AB), Julian Hynes (ON), Peter Hodson (ON), Brian Nakashima (NF), Barry Hill (NB), and staff at the PSC Office in Halifax, NS.

## APPENDICES

Appendix 8.1. Provincial and territorial addresses of agencies responsible for freshwater fisheries-related employment in Canada (listed alphabetically).

Alberta Department of Environmental Protection
Fish and Wildlife Services
Petroleum Plaza, 9945 - 108 Street
Edmonton, AB T5K 2G6

British Columbia Ministry of Environment, Lands and Parks
Fisheries Branch
Parliament Buildings
Victoria, BC V8V 1X4
and
British Columbia Ministry of Agriculture, Fisheries and Food
Aquaculture and Commercial Fisheries
808 Douglas Street
Victoria, BC V8W 2Z7

Manitoba Department of Natural Resources
Fisheries Branch
Box 20, 1495 St. James Street
Winnipeg, MB R3H 0W9

New Brunswick Department of Natural Resources and Energy
Fish and Wildlife Branch
and
New Brunswick Department of Fisheries and Aquaculture
P.O. Box 6000
Fredericton, NB E3B 5H1

Newfoundland and Labrador Department of Fisheries, Food and Agriculture
P.O. Box 8700
30 Strawberry Marsh Road
St. John's, NF A1B 4J6

Nova Scotia Department of Natural Resources
P.O. Box 698
Halifax, NS B3J 2T9

and

Nova Scotia Department of Fisheries
Aquaculture and Inland Fisheries
P.O. Box 2223
Halifax, NS B3J 3C4

Ontario Ministry of Natural Resources[1]
Human Resources Branch
Whitney Block, Queen's Park
99 Wellesley Street West
Toronto, ON M7A 1W3
or
P.O. Box 7000
Peterborough, ON K9J 8M5
and
Ontario Ministry of Environment and Energy
Human Resources Branch
40 St. Clair Avenue West
Toronto, ON M4V 1M2

Prince Edward Island Department of Agriculture, Fisheries and Forestry
P.O. Box 2000
Charlottetown, PE C1A 7N8

Québec Ministère de l'Environnement et de la Faune
Direction des ressources humaines
3900, rue Marly, 3° étage
Sainte-Foy, QC G1X 4E4

Saskatchewan Department of Environment and Resource Management
Fisheries Branch
3211 Albert Street
Regina, SK S4S 5W6

Yukon Department of Renewable Resources
Fish and Wildlife Branch
Box 2703
Whitehorse, YT Y1A 2C6

1 Fisheries biologists within OMNR are presently spread among a number of separate branches, some in different divisions.

Appendix 8.2. Locations of Department of Fisheries and Oceans regional administrative headquarters in Canada.

| Region Served | Address |
|---|---|
| Capital Region and Headquarters | Centennial Towers<br>200 Kent Street<br>Ottawa, ON K1A 0E6 |
| Newfoundland Region | P.O. Box 5667<br>St. John's, NF A1C 5X1 |
| Scotia-Fundy Region | Maritimes Centre<br>17th Floor (South)<br>P.O. Box 550<br>Halifax, NS B3J 2S7 |
| Gulf Region | P.O. Box 5030<br>Moncton, NB E1C 9B6 |
| Québec Region | C.P. 15,500<br>Québec, QC G1K 7Y7 |
| Central and Arctic Region | 501 University Crescent[1]<br>Winnipeg, MB R3T 2N6 |
| Pacific Region | Suite 408, Harbour Centre<br>555 West Hastings Street<br>Vancouver, BC V6B 5G3 |

[1] Location of administrative center pending reorganization.

Appendix 8.3. Locations of Department of Fisheries and Oceans major research establishments in Canada.

| Name and address | Area of responsibility |
| --- | --- |
| Northwest Atlantic Fisheries Center<br>P.O. Box 5667<br>St. John's, NF A1C 5X1 | Marine and Anadromous fisheries off Newfoundland and Labrador |
| Gulf Region<br>P.O. Box 5030<br>Moncton, NB E1C 9B6 | Marine and anadromous fisheries in southern Gulf of St. Lawrence |
| St. Andrews Biological Station<br>St. Andrews, NB E0G 2X0 | Marine and anadromous fisheries on the east coast and aquaculture |
| Halifax Fisheries Research Laboratory<br>1707 Lower Water Street<br>Halifax, NS B3J 2S7 | Marine and anadromous on the east coast |
| Bedford Institute of Oceanography<br>P.O. Box 1006<br>Dartmouth, NS B2Y 4A2 | Marine and anadromous fisheries on the east coast and arctic |
| Institut Maurice-Lamontagne<br>C.P. 1,000<br>850 Route de la Mer<br>Mont-Joli, QC G5H 3Z4 | Marine and anadromous fisheries of the arctic and Gulf of St. Lawrence |
| Bayfield Institute[1]<br>867 Lakeshore Road<br>P.O. Box 5050<br>Burlington, ON L7R 4A6 | All waters of the Great Lakes |
| Freshwater Institute[2]<br>501 University Crescent<br>Winnipeg, MB R3T 2N6 | Inland waters other than the Great Lakes, with focus in prairie provinces and territories |

| | |
|---|---|
| West Vancouver Laboratory<br>4160 Marine Drive<br>West Vancouver, BC V7V 1N6 | Marine and anadromous fisheries on the west coast and aquaculture |
| Pacific Biological Station<br>Hammond Bay Road<br>Nanaimo, BC V9R 5K6 | Marine and anadromous fisheries on the west coast |

1 After reorganization, research component may become part of the National Water Research Institute (DOE), same address.

2 After reorganization, research component may become part of the National Hydrology Research Institute (DOE), 11 Innovation Boulevard, Saskatoon, SK S7N 2X8.

Appendix 8.4. Provincial or regional addresses of federal Public Service Commission offices.

| Province/Region | Address |
|---|---|
| Newfoundland | Cormack Building, Suite 302<br>2 Steers Cove<br>St. John's, NF A1C 5X8 |
| Prince Edward Island | 119 Kent Street, Suite 420<br>Royal Trust Tower<br>Confederation Court Complex<br>Charlottetown, PE C1A 1N3 |
| Nova Scotia | 1557 Hollis Street<br>P.O. Box 1664, Halifax CRO<br>Halifax, NS B3J 3V3 |
| New Brunswick | 777 Main Street, 7th Floor<br>Moncton, NB E1C 1E9 |

| | |
|---|---|
| Québec | Room 205, Place Sillery<br>1126 Chemin St. Louis<br>Sillery, QC G1S 1E5 |
| Montréal | 200 René-Lévesque Blvd. West<br>Guy-Favreau Complex, West Tower, 8th Floor<br>Montréal, QC H2Z 1X4 |
| National Capital and Eastern Ontario | 66 Slater Street<br>Ottawa, ON K1A 0M7 |
| Central and Southern Ontario | Dominion Public Building<br>1 Front Street West, 3rd Floor<br>Toronto, ON M5J 2R5 |
| Manitoba | 200 - 344 Edmonton Street<br>Winnipeg, MB R3B 2L4 |
| Saskatchewan | 400 - 1975 Scarth Street<br>Regina, SK S4P 2H1 |
| Alberta and Northwest Territories | 9700 Jasper Avenue, Suite 830<br>Edmonton, AB T5J 4G3 |
| Vancouver | 757 West Hastings Street<br>Suite 210<br>Vancouver, BC V6C 3M2 |
| Victoria | 224 - 816 Government Street<br>Victoria, BC V8W 1Y9 |
| Yukon | Suite 400, 300 Main Street<br>Whitehorse, YT Y1A 2B5 |

Chapter 9

# Academic Positions in Fisheries Science

BRIAN R. MURPHY AND F. JOSEPH MARGRAF

## INTRODUCTION

Your research is completed, your dissertation is in close-to-final draft, and your defense is scheduled. Now you start to wonder, "Where do I go from here?" Many people find the idea of an academic position at a college, university, or National Biological Service Cooperative Research Unit to be attractive. Professors in these types of positions typically divide their time between teaching, advising students, university committees and administration, seeking research funds, conducting research, and writing scientific and academic publications. Time spent on each of these tasks varies widely, depending on the type of academic position and institution. Cooperative Research Units offer Federal positions at some universities, with a heavy emphasis on research and graduate training. Talk with professors about their jobs to help you decide if this is the career you wish to pursue. If the life of a professor appeals to you, then it's time to start a search for a suitable academic position.

Unlike some types of positions in private industry and government, the shotgun approach to finding employment (flooding the world with your resume) rarely works in academia. In these days of tight budgets and tightly controlled hiring practices, colleges and universities can only hire new faculty to fill specifically advertised vacancies. Watch for vacancies in the *AFS Jobs Bulletin, Science,* and *The Chronicle of Higher Education*; or posted on bulletin boards when your department receives direct mail announcements. Many openings are "announced" by word of mouth before advertisements even appear. Throughout your graduate career you should have been building a network of colleagues in fisheries, both at your own and at other institutions. Make sure that your colleagues know that you are seeking an academic position so that they can refer appropriate announcements to you.

## COLLECTING BACKGROUND INFORMATION

Once you've identified an open position that you may be interested in, you should begin to collect as much information as possible about the university, program, and the faculty in the department to which you will apply. Start at your university library's reference department. Browse through the undergraduate and graduate catalogs from the candidate school. These can be a good source of general information about the school in question, and the features of the surrounding area. Look at the degrees and options they offer in your area, and the courses that they teach. Note the names of faculty affiliated with the program to which you will apply. Next consult the *Peterson Guide to Graduate Programs in Biological, Agricultural, and Health Sciences*. This guide does not list every program in the country, but for many programs it provides information on faculty, their educational backgrounds, and their research interests. If you want further information on research, check faculty listings in *American Men and Women of Science, National Faculty Directory,* or various indexes and abstracting services (e.g., *Science Citation Index, Biological Abstracts, Zoological Record,* or *Aquatic Sciences and Fisheries Abstracts*). For information on faculty salaries at universities check the American Association of University Professors' annual salary survey, which is published periodically in *The Chronicle of Higher Education*. *Places Rated Almanac* is an excellent source of information regarding the cost of living, school system ratings, crime rates, climate, recreational opportunities, and other general information for many metropolitan areas of the USA. Local newspapers also can be an excellent source of information regarding local issues and the cost of living in the areas in question.

Feel free to contact personal acquaintances that you may know in the department to which you will be applying. This can be a good way to get the "inside scoop" on exactly what the faculty hope to get in a new colleague, but don't expect your acquaintances to provide information that they couldn't provide to every caller. If you don't know anybody in the department, don't hesitate to contact the search committee chairperson for answers to questions you might have about the position or the program, but don't be a pest throughout the search process.

## THE APPLICATION PROCESS

### What to Send

Read the job announcement carefully, and follow all directions for submitting an application. Send only those materials requested, which generally will be a statement of interest in the position (i.e., a cover letter), a complete resume, and a list of persons who have agreed to serve as professional references (including their phone numbers). Do not have references send letters of recommendation immediately unless the job announcement specifically states that you should do so. Often the search committee will make a first cut of applicants before requesting letters of recommendation. Typically, the search committee is faced with going through dozens or even hundreds of applications; try to make their job as easy as possible. Skip sending a large packet of reprints that nobody will look through, and never send more than just a few of your best publications if you were instructed to send reprints. A brief statement of teaching and research interests often is requested, or could be added at your discretion. All materials that you send should be complete yet concise, well organized, and well written. Make certain that you check for typographical errors and misspellings several times before you mail your application.

### Your Resume

Your resume should be in an easy-to-read format that provides all critical information. Chapter 2 provides a detailed discussion of resumes, but there are some important points to consider for the academic application. Customize your resume for each position to which you apply. Think about the many varied tasks that faculty members perform, and show how you have experience and accomplishments that demonstrate your potential for success in the position in question. Don't try to pad your publication list with all types of reports, presentations, abstracts, and other non-refereed items. Show refereed publications in a clearly labeled section, and either put other types of publications in their own sections or in a general "Other Publications" section. List all co-authors on all publications, in the proper order of authorship. Limit listings of personal data and hobbies to only those things that will have some bearing on your qualifications for the job. Avoid flamboyant paper colors, fonts, and graphics for your resume. Don't let your experience and accomplishments be overshadowed by aggressive marketing ploys that may turn off some people.

## THE INTERVIEW

The search committee will sort through all applications submitted, and generally select three or four candidates for interviews. You may be asked to interview with very little advanced notice, so be preparing for an interview before you actually are contacted. You can be sure that you will be asked to present a seminar during any interview, so begin preparations early.

### Meeting People

The impression you make on people when you first meet them might turn out to be one of the deciding factors in their selection process, so do your best to create a good impression. Dress appropriately (coat and tie for men, equivalent level of formality for women), and try to be friendly but not pushy. Be aware that your body language may sabotage an otherwise successful interview. Shake hands firmly with people you meet; maintain confident, but not glaring, eye contact; and be sure to use natural, erect posture both in sitting and standing. Expect to be scheduled to meet with the search committee, various faculty members, and any number of university administrators from the dean to possibly the president. Be prepared to carry the conversation in some of these meetings, as busy faculty and administrators may not have taken time to reread your resume before meeting with you. Expect to be asked to talk about your background and interests. Be prepared for questions like "Why do you want to come here?", "What do you perceive as your greatest strengths and weaknesses?", or "What is the first thing you would do if you were hired?" Keep your answers positive, as you don't want to come across as a complainer or as hard to please. Be prepared to discuss the types of classes you would be interested in teaching, and research interests that you would plan to pursue. Be ready with questions of your own about the university and the local area to keep the conversation from lagging. You also might be scheduled to meet graduate and undergraduate students, departmental staff, and even a local real estate agent; if not, ask if these meetings might be possible, as each of these can give you valuable insight on the university and the local community.

**The Interview Seminar**

Undoubtedly you will be asked to present a seminar for faculty and students. This is not just an easy way for faculty to learn about your research interests, but is also a test of your communications skills and potential as a classroom teacher. Gauge the audience. Know what to expect, and be sure to have something in your seminar for viewers at every level. Don't try to amaze everybody with your technological or statistical prowess. Present a balanced seminar that shows both depth and breadth in your abilities. Be sure that your slides are free of errors, and stay within your allotted time. Anticipate questions that are likely to be asked. Don't be surprised if you also are asked to present a guest lecture in an undergraduate class. This tactic is becoming more common to give the search committee a solid appraisal of your teaching skills.

**The Exit Interview**

Often, your last meeting will be with the department head or chairperson. This will be your chance to discuss your expectations and those of the university, if these have not come up already. Be sure you have a clear understanding of what space, equipment, and start-up funding will likely be available for the successful candidate. Also clarify the university's and department's expectations for teaching load, accomplishments for tenure, etc. If you are not asked about salary expectations, bring up the subject. You should expect a salary level near the median of faculty at the same rank in which you would start. Be sure you understand whether the position is a 12-month appointment or something less. The interview will likely close with a discussion of how you view the program and the potential position. Be prepared to talk about potential research collaborations that you may have discovered during the interview.

## THE COOPERATIVE RESEARCH UNIT PROGRAM

Academic jobs are also available in the Cooperative Research Unit Program. These are Federal positions stationed on a university campus, with the university granting full academic and graduate faculty status. Cooperative Unit faculty teach advanced courses, advise graduate students, perform or lead scholarly research, and in most other ways perform duties as other faculty members.

**Staffing at Coop Units**

A Unit is usually staffed with three professional biologists employed by NBS. Because they must meet the university's requirements for appointments as faculty members approved to advise graduate students, they must hold a Ph.D. in an appropriate field of study. The Unit Leader may be trained in either a fisheries or wildlife discipline, whichever is deemed appropriate by the Cooperators to meet the needs of the Unit. The Unit Leader is a senior scientist, who has usually served as an Assistant Unit Leader somewhere in the Unit Program. In addition to his or her scientific duties toward accomplishment of the Unit's mission, the Leader is administratively responsible for the activities of the Unit. There are two Assistant Unit Leaders, one trained in fisheries and one in wildlife, whose primary duties are directed to accomplishment of the Unit's mission. Assistant Unit Leaders may be at any stage in their professional careers from entry-level to senior scientists. New or recent Ph.D. graduates can expect to compete for positions as Assistant Unit Leaders.

**Applying for Unit Jobs**

Openings for Assistant Unit Leader positions are usually advertised in much the same way as university faculty positions. The application process, however, is different from that for a faculty position because Unit staff are Federal Civil Service employees. The applicant must submit a completed SF-171 form to the appropriate government Personnel Office (as shown on the job announcement) and should be sure to completely address the ranking Factors of Knowledge, Abilities, Skills, and Other Characteristics listed on the job announcement (see Chapter 6). The Personnel Office will pre-screen the applicants and forward the certified list of eligible candidates to the university and state cooperators. From this list, the cooperators will choose their Assistant Unit Leader, using an interview process much the same as that for a faculty member. The individual is chosen by consensus of all the Unit Cooperators.

## CONCLUSION

Preparation and planning are the keys to a successful academic job search. Plan your graduate program to provide the experiences that will make you an attractive applicant. Do your homework before applying and before interviewing. Going to an interview well prepared will allow you to relax, be yourself, and make a good impression.

Chapter 10

# Working For Non-Governmental Organizations

ANDREW J. LOFTUS

For the most part, people entering the fisheries and wildlife professions do so because of an interest in working in the outdoors or working directly with animals or animal populations. However, it doesn't take long to realize that the adage of fisheries management being "90% people and 10% fish" is certainly true. Whether entering the public sector or private industry, fisheries professionals will all have to rely on, and report to, constituency groups in order to be successful in achieving management, educational, or research goals.

Many constituency groups are represented by organizations designed to promote their interests. These organizations are commonly referred to as non-governmental organizations, or NGOs. Working for a NGO provides a perspective of fisheries management that cannot generally be fully gained through formal education. There are a growing number of organizations which cater to an increasingly specialized constituency. These organizations may rely on financial support from individuals to carry out their programs or they may be supported by companies and entire industries which rely on natural resources (for example timbering, sport fishing, commercial fishing etc). Additionally, some environmental foundations are supported through endowments or private trusts.

## EDUCATION

Most college fisheries and wildlife science and management curricula emphasize biology, taxonomics, fisheries/wildlife management and statistics while requiring minimal courses in human dimensions, economics, public relations, public speaking or writing. However, when working with NGOs, it is generally helpful to have a balanced education in all of these areas. Often, little beyond what is required as part of the university's general education program in the way of social science is required for fisheries and wildlife students. Although some schools are tailoring social science or human dimensions courses specifically for fisheries and wildlife students, these courses still are not prevalent in most curricula. However, students who may be planning for employment with NGOs should have a strong understanding of fisheries management supplemented with non-traditional courses such as those mentioned above. In addition, a more thorough knowledge with economics, political science, public speaking, journalism (non-scientific writing) and education can be valuable depending on the choice of career paths.

## TYPES OF JOBS

The jobs available to those trained in fisheries and wildlife fields will vary tremendously depending on the NGO for which you are working. However, there are some general classifications that can be made. Keep in mind that NGOs typically employ people from a diversity of disciplines, career focuses, and backgrounds. The classifications listed below highlight only those employment possibilities for which natural resources professionals may be best suited.

### Resource Issue Analysts

From a resource perspective, many NGOs work directly with government agencies to develop and implement management programs which benefit the resources of concern to the group. When presenting information before government agencies or policy makers, it is extremely important to present credible testimony. Therefore, many opportunities exist for professionals with an understanding of management principles and a general understanding of how government management agencies operate. Most NGOs rely on professionals who can evaluate the impact of proposed regulatory actions and, if needed, develop alternatives to proposed actions. For these types of positions, education in both resource management and experience working with government agencies is highly valuable.

### Government Relations

Government relations is another important component of NGO programs. The objective of government relations programs is to interact with policy makers and agencies to ensure government support and regulations which benefit the goals of the organization.

Generally, fisheries and wildlife professionals, unless already established with solid credentials in government relations, would start in a position supporting a government relations department (such as a resource analyst position) before moving full time into this. Commonly, experienced fisheries and wildlife professionals, having worked their way through the administrative ranks of state or federal management agencies, may enter into government relations positions with a NGO.

Government relations requires strong skills in interpersonal relations, public speaking, and writing. Those entering this field must be flexible, be able to work on diverse issues of concern to the NGO, and have a thorough knowledge of legislative and regulatory processes.

**Information and Education**

Education is an important component of NGO programs. A goal of most organizations is the education of its members and the public about the mission of their organization. Strong skills in communication are required for these types of positions. Students and professionals with background as naturalists may find these jobs highly satisfactory.

**Volunteer Coordinator/Grassroots Activism**

Many organizations rely on a network of members or activists to implement their programs at local levels. The ability to access and communicate with these grassroots individuals requires a thorough understanding of resource issues and skills to convey complex issues to a diverse membership, mostly in a nontechnical manner.

**Biologists/Managers**

Some NGOs are focussed primarily on "hands-on" management of land and water resources. Often these groups are involved with the actual acquisition of land, easements, or property rights. As such, the groups require professional biologists to manage the resources.

**Foundation support**

Some NGOs have established foundations or trusts to fund projects which further the objectives of their organization. These may include funding projects which are designed to enhance the resources. As such, organizations may require individuals with strong resource backgrounds who can evaluate both the technical merits and the cost/benefits of proposals submitted to them for funding consideration and who can interact with resource agencies and groups in a technical capacity.

## EMPLOYMENT

In general, employment with NGOs may not be as secure as employment with government agencies. Initial employment opportunities with groups may require accepting a position which is funded on a limited basis or for a specific project. Such opportunities though can lead to networking and more permanent employment in the future. In general, NGOs are not restricted to similar hiring procedures as government agencies and thus are able to hire individuals when needed, even if an individual may not have all of the formal training called for in the position.

### Internships

If you are not certain that a career with a NGO is for you, a good way to explore this option is by accepting a temporary internship. Many groups offer these types of limited employment in conjunction with a specialized project which the group is undertaking. Additionally, internships offer a means to network with others in the field. In many cases, the smaller staff size and less bureaucracy of NGOs mean that individuals can play a role in a number of issues and jobs that may be outside of the individual's area of training or expertise. These opportunities are invaluable for helping to identify areas of interest or disinterest and developing work experience in diverse areas. Internships may be paid or unpaid, with both types sometimes being offered within the same organization.

### Volunteer

Volunteering to do work for a NGO offers both you and the group an opportunity to evaluate one another. Volunteering can give you experience in an organization's actual workings, goals, and employment opportunities which may be available to you. Furthermore, volunteering provides you the opportunity to showcase your talents with a group and gives you an inside track on potential future openings. In many cases, for those who are already employed and are unsure whether they want to make a move to a NGO, volunteering their time to the organization for a few hours a week can provide additional insight to assist in their decision.

### Intergovernmental Personnel Agreements (IPAs) or Cooperative Agreements

The federal government as well as some state governments allow certain employees the opportunity to work with non-governmental organizations without losing civil service status in the government. These IPAs vary in form, but commonly the government agency will pay the salary of the individual with the organization providing office space, support staff, and other services to the employee. The length of IPAs can range from a few months to two or more years, depending on the circumstances. An IPA is an excellent way for the government and private sectors to share resources and draw upon each other's strengths. For the individual, it offers a secure way to experience the workings of an organization and gain a greater appreciation for the roles and mission of the organization. Check with your current agency directors or personnel office to see if your agency would be amenable to participating in IPAs.

## GETTING STARTED

Finding out about an organization may not be as difficult as it seems, and is a necessary step to ensure a constructive employment experience. There are several ways through which to investigate an organization.

### Personal Visits

There cannot be enough said for actually visiting an organization. Meeting with the professionals employed with the organization will give you a feel for what the attitude and approach of the organization may be like. Additionally, a personal visit will leave an impression with the organization's leaders that you are truly interested in working for them.

**Professional Meetings**

Professional meetings provide an excellent forum in which to network. Some NGOs may even have a display or booth at the meeting where you can speak with them about their program and learn about prospective job openings. Often times, NGOs host meetings or participate in the presentations at the meetings, giving individuals an opportunity to find out more about their activities.

**Publications**

Most organizations publish newsletters or other literature which carry a wealth of information about the organization. Newsletters will familiarize you with their programs before you interview or speak with someone about a position. When hiring new employees, most organizations look for people with interests in the issues of the organization- and what better place to learn about these issues than from the organization's publications. Contact the organizations you may be interested in to inquire about being added to their mailing list.

Additionally, the "Conservation Directory," published annually by the National Wildlife Federation, provides a description of most major NGOs in the U.S. This directory will not only provide a brief description of the organization, but also contains contacts, phone numbers, and other information.

**Annual Business Meetings and Other Events**

Most NGOs will hold business meetings at least once per year for their membership. The structure of these meetings will of course vary with each organization. Business meetings of larger NGOs often incorporate seminars and other outreach activities during their meetings, which provide ideal opportunities for an individual to learn more about the fundamentals of the organization. Additionally, as part of their membership and outreach programs, several NGOs host a variety of events throughout the year ranging from instructional workshops to trade shows. These opportunities may be valuable for individuals desiring to interact with the membership and officers of the organization. A word of caution: Since some of these events may be restricted to members only, before making a special trip to attend an event, be sure to inquire as to whether the event is open to the general public and the registration process/costs involved.

**Washington D.C./State Capitols**

Many NGOs have offices in state or federal capitols, whether they are heavily oriented toward government relations or not. This provides the organizations with a centralized location where they can interact with other NGOs and government agencies on a regular basis. Check listings for organizations in the state in which you wish to work.

**Local Organizations**

There are a tremendous number of smaller organizations which are formed to focus on specific issues or resources of local concern. Opportunities in these organizations may be limited due to their smaller staff sizes and budgets. However, if you have the background or expertise in issues in which they are interested, these organizations may be inclined to develop sources of funding or otherwise acquire your talents. Although more difficult to locate than larger organizations, they can be found often by contacting larger "alliances" of organizations which are commonly formed in state or national capitols.

## CONCLUSION

Employment with NGOs can be extremely rewarding and challenging. NGOs can offer that "new opportunity" that many professionals look for at some time during their career. When evaluating NGOs for potential employment, remember that NGOs are generally accountable to a Board of Directors and their membership base, so choose a group whose mission best coincides with your personal goals.

BAIT

Chapter 11

# Employment in Aquaculture[1]

NICK C. PARKER AND ARDEN J. TRANDAHL

Aquaculture is defined as the production of aquatic species -- vertebrates, invertebrates, and plants. Although many people identify aquaculture with fish farming and even more narrowly as the production of fish in fresh waters, aquaculture is similar to agriculture and produces a wide variety of food fish, hobby fish, bait fish, ornamental plants, shrimp, mollusks, and algae. Today, about 20% by weight and 30% by value of all fish and fisheries products are produced by aquaculturists. As the catch from global fisheries declines, the importance of aquaculture products is expected to expand. Job opportunities in aquaculture are wide ranging and employ the skills and training of all levels from high school graduates to individuals with highly specialized doctoral degrees. This chapter presents some of the job opportunities in aquaculture with the skills and training required for those jobs.

## COMMERCIAL PRODUCERS

Fish farms may be privately owned or owned by large corporations. Depending upon size, fish farms employ individuals with varied levels of formal training and experience. Frequently, the owner or manager of a fish farm will be a professionally trained aquaculturist with a bachelor's, master's or doctoral degree in fisheries or aquaculture. If not, management will have access to such an individual. Expertise required of personnel working on a fish farm includes the ability to spawn, feed, and care for fish of all sizes and ages in marine, freshwater, warmwater and coldwater environments. At least some of the employees must be trained to monitor water quality using specialized equipment to measure oxygen, ammonia, pH, etc., and to monitor fish health by routinely using microscopes to

---

[1]Publication No. T-9-705 of the College of Agricultural Sciences and Natural Resources, Texas Tech University.

examine fish and microbiological techniques to identify fish pathogens.

Fish-farm personnel may be employed to culture aquatic organisms such as rotifers and other invertebrates as food for larval fish. Still others are involved in construction of ponds, installation and maintenance of pumps and other equipment. Commercial fish producers also employ personnel to harvest and grade aquatic species for restocking or to pack and ship them for sale. For some of the simpler tasks, graduates of two-year technical schools are often highly sought as potential employees by commercial fish farms. Individuals with formal classroom training in aquaculture often secure their first experience through internships or other on-site programs providing hands-on experience. Job opportunities exist for individuals with undergraduate and graduate level degrees; however, the number of jobs available is inversely proportional to the level of training.

## GOVERNMENT AGENCIES

Some of the earliest aquaculturists were employed in government hatcheries to propagate and stock cultured finfish and marine invertebrates. Jobs for aquaculturists still exist at many levels of government.

Many local, state, and federal agencies with technical assistance programs or regulatory responsibilities employ aquaculturists. Individuals with an understanding of aquaculture or with special training ranging from associate to doctoral degrees are employed by public agencies to provide assistance to the industry through extension offices and diagnostic laboratories. Other aquculturists have been employed to assist with financial planning, insurance, marketing, etc. Agencies which employ aquaculturists at the national level include the U.S. Department of Agriculture, U.S. Fish and Wildlife Service, U.S. National Biological Service, National Marine Fisheries Service, Bureau of Indian Affairs, and the Department of Defense. At the state level, aquaculturists are employed by the department of agriculture, the department of natural resources, the water quality agencies, and state extension offices. When aquaculturists are hired at the local government level, it is usually in association with a state or federal program.

Research opportunities in laboratories operated by various

government agencies provide challenging employment opportunities for individuals with a multitude of specialties related to aquaculture. Chemists, biologists, physiologists, endocrinologists, veterinarians, engineers, nutritionists, and numerous other specialists are employed in laboratories in support of aquaculture programs. Most government jobs require specialized training at the level of a bachelor's degree, while others require a master's or doctoral degree.

## UNIVERSITIES

Private and public universities employ individuals with skills in aquaculture or supportive of aquaculture. Probably every Land-Grant university and all college programs associated with Sea-Grant universities employ faculty and research personnel with knowledge and expertise in aquaculture. Job opportunities range from administrative positions and professors to technicians and graduate students. Programs are housed in the departments of agriculture, biology, engineering, and fisheries. Daily activities may include classroom instruction, research in laboratories (on campus or at off-campus sites) and diagnostic and extension services. Most job opportunities in universities are for individuals with graduate level training -- master of science and doctoral degrees.

## SUPPORT SERVICES

A wide variety of job opportunities for individuals at all levels of training exist in support of the aquaculture industry. Individuals with training ranging from two-year associate degrees through doctoral degrees in fisheries and in aquaculture often are employed to assist in developing and marketing supplies and services for the aquaculture industry. Opportunities exist for individuals with master and doctoral degrees in the development and marketing of chemicals including drugs, hormones, herbicides, anesthetics, etc. Hardware manufacturers and vendors employ individuals with bachelor and master degrees to supply tanks, aquariums, pumps, seines, nets, analytical equipment, blowers, buildings, pond liners, aerators, and other equipment and supplies within the aquaculture industry.

Consulting and engineering firms hire aquaculturists to assist in design and operation of facilities, to solve specific problems within the industry, and to provide timely guidance to owners and investors of aquaculture firms. Fish processing plants, wholesalers, and marketing and advertising firms hire individuals knowledgeable of the aquaculture industry. Every manufacturer of fish feeds employs at least one individual with training in aquaculture. Some feed manufacturers employ aquaculturists to conduct in-house research and to provide support within the industry through extension activities. Some aquaculturists have moved into the retail business and work in pet shops to sell ornamental fish, aquatic plants, and supplies. Others are employed, principally in California and Florida, to inspect, package or repackage live fish for interstate or international shipment.

## SUMMARY

Aquaculture is a relatively new and growing field with a wide array of job opportunities. Many of the opportunities for employment are developed and shaped by the skills and traits of the successful job applicants. Worldwide, no area of agriculture is expanding as rapidly as aquaculture. The future for employment in aquaculture and its related fields looks very bright.

## ACKNOWLEDGMENTS

The authors appreciate the assistance of R. Patiño, National Biological Service; S. Demarais and J. Winter, Texas Tech University, for their reviews and editorial assistance.

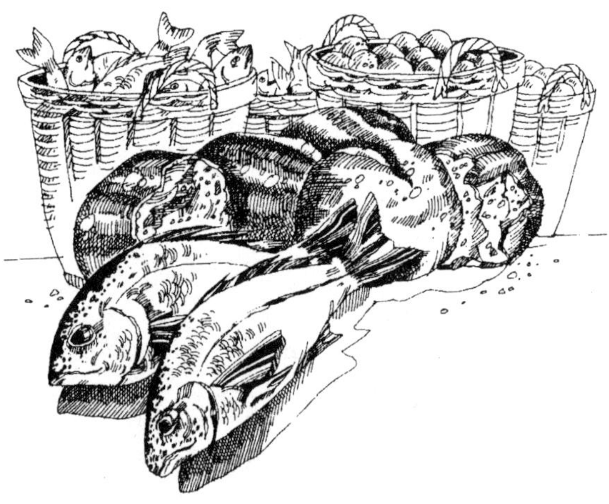

Chapter 12

# Private Consulting

DONALD MACDONALD AND FORREST OLSON

## OPPORTUNITIES

Depending on your perspective, employment opportunities provided by consulting firms represent either the chance of a lifetime or the bane of the universe. Regardless of your point of view, however, private consulting firms, both small and large, are providing an expanding share of the fisheries employment market and should be considered by serious job hunters. Recent surveys indicate that about ten percent of professional fisheries biologists are employed by consulting firms. Small firms typically employ one to ten people and are often highly specialized; therefore, these types of firms tend to target relatively small and focused projects. However, by developing cooperative associations with other independents or with larger firms, small firms can acquire the expertise necessary to bid on bigger multidisciplinary projects.

Larger firms, which typically consist of engineers and planners as well as scientists, tend to work on larger multidisciplinary projects. As such, fisheries biologists in large firms have an opportunity to work on a variety of projects, of which fisheries issues may be only one component. Some firms specialize in bioengineering projects such as hatchery designs and fish passage facilities, but most deal with projects that involve fish habitat. In many cases the consultant biologist plays the role of liaison between his client, often a developer, and the resource agencies charged with protecting habitat. This role can be especially rewarding to those interested in the "people" aspect of work rather than merely the technical details.

Consulting firms provide a number of employment opportunities for students, recent graduates, and established professionals. Undergraduates are attractive to independent consulting firms because a number of government and university programs that subsidize student wages are available. Cooperative education and other programs encourage firms to hire students for 4 or 8-month work terms. Experience gained during these work terms is invaluable for securing full-time employment after graduation. Many independent consulting companies also hire recent graduates on a part-time or temporary basis, thereby providing new professionals with excellent opportunities to gain practical experience. Recent graduates or established professionals may also develop their own small firms, which gives highly motivated individuals a chance to excel.

## QUALIFICATIONS

Consulting firms tend to operate with tight deadlines and limited budgets. Therefore, employers are often looking for people who can contribute immediately to the firm. When evaluating undergraduates, employers generally look for good grades, some work experience and computer skills, flexibility, eagerness, and a willingness to work. In addition to these general qualifications, a university degree, relevant work experience, proven communication skills, and the ability to work on a team are important qualifications. Individuals who think of themselves first and the team second need not apply!!

Consultants employ about equal numbers of people with bachelor and master degrees, but in terms of getting a job, a master's degree will definitely be an advantage. A degree in fisheries should not be viewed as a necessity. Other related fields such as ecology, watershed science, and hydrology are also important to firms working in the fisheries field. Such firms are usually involved in terrestrial studies too, including wetlands, wildlife, and vegetation. A varied background in fisheries, coupled with other related fields, is always attractive to a consulting company.

Whereas a college degree is the foot in the door, relevant work experience is what counts most in getting hired. What is relevant experience can be quite variable, depending upon the type of clients and projects in which the consulting firm specializes. Most private firms work on projects involving habitat alteration issues rather than fisheries management issues (this being largely in the public domain). More and more consulting firms are becoming actively involved in ecosystem studies, watershed analyses, and cumulative impact assessments, and thus any experience or training in these areas would be considered an asset when looking for a job.

One of the best ways to get the work experience required to

gain employment with a consulting firm is to work for government, industry, or academic institutions as an undergraduate. Alternatively, volunteering in community-based enhancement or habitat rehabilitation projects often provides important experience and demonstrates initiative and devotion to the field. Employers also look for individuals who are serious about their profession. Show your commitment by attending professional meetings, preparing presentations, and publishing papers. All of these activities will increase your confidence, hone your skills, and make you more attractive to potential employers.

Communication skills are always high on the list of qualifications sought by consulting firms. The primary work products of consultants are reports, and even the best scientist will not be an effective consultant without the ability to convey information and thoughts in writing.

## GETTING HIRED

The job market in the fisheries and aquatic resource management field is very competitive and becoming more so every year. To get hired, you must stand out from the rest of the crowd. The first step toward getting hired is to prepare a resume and cover letter (see Chapter 2). These introductory tools provide you with a chance to demonstrate your written communication skills to prospective employers. In the cover letter you should highlight your knowledge, experience, and capabilities as they apply to the position that is posted or one that you think may become available. Definitely highlight your relevant work experience and volunteer efforts. While an impeccable resume and cover letter will not guarantee you a job, sloppy resumes and/or cover letters are virtually always destined for the wastepaper basket. Have your resume and cover letter reviewed before submitting them to potential employers.

The "fill-in-the-blank" canned resumes and cover letters that seem to be popular now can be spotted a mile away and can be a real turnoff. Do yourself and your prospective employer a favor and write a genuine resume and cover letter that reflect your true self.

Consulting firms usually do not prepare and distribute advertisements for a position. If they are looking for someone with special experience or capabilities, they usually first spread the word of the position opening among professional contacts in agencies, universities, other firms, or professional organizations. For entry-level positions they often rely on their own files of recently submitted "unsolicited" resumes. Therefore, it is important for you to contact firms frequently so that they think of you when they need help. Make an effort to visit the office and get more information about the firm and its activities. Show an interest. Offer information about yourself. Demonstrate a sincere willingness to work and to work hard. In many ways this approach is superior to waiting to be called for an interview because it is informal and neither party feels pressured into making a decision. Nonetheless, the employer is likely to remember you and call you when assistance on a project is required. Also, you will gain important information for deciding if you really want to work for the firm.

Whether you schedule an informal discussion with the company principal or are called to a formal interview, it is essential for you to make a good first impression on the prospective employer. Therefore, it is useful to treat the interview like an exam. Study for it. Try to become familiar with recent studies in the field, especially those with which the firm may be involved. Inquire about regional fisheries and environmental programs and issues. It is this type of effort that will set you apart from other applicants-and get you a job. Do not be surprised to be offered short-term or project-related work in the beginning. Private firms rarely make a long-term commitment without *knowing* that you can make a significant contribution to the team. Even if you are not offered longer term employment with the firm, this experience will prepare you for employment elsewhere. Either way, you are a winner.

## ADVANTAGES OF WORKING FOR A CONSULTING FIRM

There are several advantages to working for a consulting firm. First, it offers you an opportunity to excel. If you demonstrate a willingness to accept responsibility and have the capabilities needed to participate effectively on the team, you will succeed. In addition, this type of employment offers a variety of work experiences, limitless opportunities to learn, and some ability to select projects of interest. Moreover, this field provides an opportunity to be involved in topical issues and to make a difference in the management of the resource. Over the long term, association with a private firm or developing your own company will also provide you with a chance to succeed financially as well.

## DISADVANTAGES OF WORKING FOR A CONSULTING FIRM

Perhaps the biggest downside to working for independent consulting firms is the lack of job security relative to government jobs. While business has generally been pretty good over the past 5 years, there are no guarantees beyond the contracts that are already signed. In addition, salaries are typically low over the first few years. However, salaries generally improve significantly as you become established in the field. Overtime is usually compensated on an hour-for-hour basis; so do not expect to augment your salary with all the weekends and holidays that you will be working. Importantly, retirement, medical, and dental plans are often not provided in small firms, and this can represent an added strain on your financial resources.

If you enjoy long, relaxing vacations, do not even consider working for a consulting firm. Your schedule is frequently driven by tight deadlines, and unexpected projects commonly destroy the best laid vacation plans. However, if you can live with the uncertainty and pressures, employment in this type of firm provides challenges and job satisfaction that are unparalleled in other sectors.

# Chapter 13

# Professionalism in Fisheries Science

PAUL BROUHA AND ROBERT L. KENDALL

## MAINTAINING SOCIETY INVOLVEMENT

Membership in the American Fisheries Society (AFS) does not automatically make one a professional, but it is a good indicator of professionalism. Like other professional societies, AFS offers its members many opportunities for career and leadership development, for communication among peers, and for advancing common interests. The Society mobilizes the expertise of its members to address scientific and management concerns in fisheries. It pools the voices of its members to speak with strong effect on aquatic resource issues. It generates the publications, symposia, workshops, and continuing education courses that keep the fisheries profession up-to-date and effective. The Society helps maintain and improve standards of performance and public responsibility within the profession, and it encourages high-quality work through an extensive awards program. In turn, the Society works to enhance the profession's public stature and to increase public support for fisheries science and management.

To say that AFS "does" all these things is to say that its members do; AFS is simply the organization they have created to magnify their individual efforts and to improve themselves and others as scientists and stewards of fishery resources. Every fisheries professional in North America has a stake in at least some of the issues that AFS addresses; therefore, every professional ought to participate in--and help define--the Society's agenda. Its organizational diversity allows AFS to support member activities at local, regional, national, and international levels and within discipline specialties. The Society has never offered more opportunities to participate in professional activities than it does today.

The benefits of participating in AFS accrue at every step along a career track. Students--the rising generation of fishery professionals--begin to establish their network of professional relationships by attending AFS-sponsored meetings at all levels of the organization. As their research develops, they can showcase it (and themselves) to critical but supportive audiences at AFS meetings and gain valuable experiences in technical communication. By participating in Chapter affairs and serving on volunteer committees, they broaden their exposure to professional issues and encounter chances to assume responsibility and exercise leadership. Society involvement also brings them greater access to scholarships, fellowships, grants, and jobs, further aiding their growth from apprentice to professional.

Entry-level to mid-career fishery professionals find AFS membership an increasingly valuable asset. Society-sponsored meetings allow them to steadily expand their networks of colleagues and information sources. Their work can be published in a wide variety of AFS journals and books. Publications, symposia, workshops, and continuing education courses help them maintain and hone their technical edge and, as they move into supervisory positions, their managerial skills; if they cannot find what they need, the Society often can help them create it. Their qualifications can be peer-reviewed through the Society's professional certification program. If they work for a public agency in the United States or Canada, they will value the periodic AFS surveys of fisheries salaries at all levels of state, provincial, and federal government; past surveys have led to improved compensation in some instances. Whether their interests lie in technical specialties or broad resource policies, established professionals can find other AFS members--and usually an AFS subunit or committee--to help them further those interests. Perhaps most importantly, their sustained participation in AFS affairs earns them growing responsibility and stature within the Society and thus within the profession. This allows them to exert greater influence on the science-based defense and management of aquatic resources.

Experience and professional recognition are cumulative, and senior people typically have the best opportunities to affect fisheries policy at national and international levels. If they have been active in AFS--especially if they have led Chapters, Divisions, or Sections and served on important AFS committees--they understand how to mobilize Society resources to accomplish important professional and fishery objectives, thus leveraging their personal capabilities manyfold. Even if they do not choose leadership roles, senior professionals can influence AFS policies and programs through the wisdom of their counsel and the mentoring they offer younger members. For themselves, continued interactions with old and new colleagues at AFS functions keep them fresh and abreast of fishery issues. Well after their official retirement, many people remain active and productive in fisheries, in part because of the stimulation and learning they obtain from continued participation in AFS affairs.

At any level of professional development, what members can get out of AFS is directly proportional to what they put into it. Sustained involvement brings the greatest rewards.

## COMMUNICATION

Professional benefits from AFS or any source do not arise in a vacuum; they stem from human interactions, and this requires communication. Effective communication is important for career development, but it also is essential for mobilizing others to serve a common cause. Furthermore, fisheries professionals have a public responsibility to communicate their knowledge, because the public has invested in their training, will continue to invest in their work, and has entrusted some of its most valued resources to their care.

Both informal and formal communications are professionally important. The informal networking that goes on at meetings, in hallways, and by phone, fax, and mail is an efficient way to give and receive current information, and it cements the personal relationships needed for cooperative activities. As one's network grows, so does one's ability to cope with developments and influence them. The only practical limit on network size is the time available to maintain it. Professionals cannot function effectively without networks.

Formal communications have structured content and logical development; they result from, and often display, rigorous thinking about some subject. Their purposes range from coercion (e.g., fishery regulations) to persuasion (e.g., grant applications), and they embrace academic lectures, public speeches, essays, white papers, policy documents, legislative testimony, many kinds of letters and reports, and scientific papers, among others. They contain little "raw" information; rather, the information has been summarized, edited, and interpreted with the intent of influencing the behavior or thinking of some target audience. The audience is more or less anonymous (journal readers, fishing club members), different in mind-set from the communicator (students, senior agency directors), or both.

Formal communications are among the most difficult tasks facing fishery (and other) professionals because of the intellectual rigor they require and the difficulty of foreseeing audience reaction to them. But they are essential tasks nevertheless. Taking them on and doing them well will bring professional advancement. Shirking them or doing them poorly will stunt a career. Here are a few tips for effective formal communication.

- Know the principles of language. They undergird everything, spoken or written. Bad grammar, tortured syntax, and incomprehensible word strings can kill a message--and debilitate the messenger. Take more courses if necessary. Practice.
- Tell a story that will be interesting to the target audience, but do not "show off." Remind the audience why it should be interested in your message, but minimize hype. Preview your conclusions so the audience can better understand the force of your argument as it subsequently develops. Make sure the audience does not get lost as the story unfolds. Introduce information when it is needed, not before or after. Excise material that does not contribute to the story.
- Learn as much as possible about the audience in advance. Read the journal to which you will submit a manuscript, and talk to people who have dealt with the editor. Find out before you speak to a sporting club whether members fish for walleye or bass; whether they favor harvest or catch-release. Determine what the audience already knows and tailor your presentation accordingly. Analyze your experience with each audience so you will be better prepared for the same or a similar audience in the future.
- In the same vein, learn the dialect and culture of the audience: the terms and phrases people use, the words that attract or annoy them, the ideas they accept or question. Use metric units of measure if the culture demands it. Avoid erudite jargon if the audience will not comprehend it. If you challenge a community truth, make appropriate bows to the existing paradigm, make the change seem evolutionary

rather than revolutionary, and do not gloat.

- Respect the audience. Do not communicate over its head or insult its intelligence. Be honest with it. Tell it something useful. Do not waste its time.
- Accept criticisms professionally, whether they come from the audience or from peer reviewers (proxies for the audience). The majority of criticisms are provoked by bad scripts, not by inherently bad messages. Script problems can be corrected. Try to anticipate problems by consulting members of the future audience. Reviews by friends, spouse, and close colleagues can help a great deal, but realize that such people are members of your culture, not necessarily the culture of your intended audience.
- Follow instructions. If a journal editor requires four copies of a manuscript, send four--not one or two. If a session chair says "35-mm slides only," bring slides--not overheads. Meet posted deadlines. Rules are set to expedite the process of evaluating and presenting information. The people who manage that process often control which information will be presented. It is risky to annoy them with unnecessary problems.

In the fisheries profession, those who express themselves clearly and well in the language of their audience earn extra measures of respect and influence, and their careers advance accordingly. Communication skills should be learned early and honed daily. One's first manuscript submission, public speech, or job interview will show the fundamental importance of those skills.

## ETHICS

Personal honesty and respect for others is at the core of ethical systems in North America. Standards of personal ethics--learned from birth and reinforced in various ways by society--give individuals a clear sense of right and wrong and of how to interact with other people.

Respect can be given to institutions and ideas as well as to individuals. In a system of professional ethics, the profession itself is a focus of respect. For example, the American Fisheries Society's <u>Code of Practices and Ethics</u> states that "Members ... strive to preserve and enhance the dignity of the fisheries profession." In part, they accomplish this through personal honesty in reporting their work, in dealing with colleagues, employers, clients, and the public, in representing their qualifications and expertise, and in resisting coercion of their professional judgments. In part they do it by according respect to other professionals: giving credit for their work, exchanging information, restricting technical criticisms to technical forums, and maintaining confidentiality. And they do it by self-policing, encouraging good conduct in others and exposing misconduct when they encounter it.

To some extent, professional ethics (like individual ethics) are solutions to problems of security and survival, but they can embrace altruistic ideals. "First and foremost," proclaims the AFS Code, "members accept the responsibility to serve and manage aquatic resources for the benefit of those resources and of the public" (emphasis added). This level of respect for human and nonhuman systems has not reached the sophistication of Aldo Leopold's land ethic, but it is closer to that ethic than it used to be. Only in the last quarter century have law and practice begun to codify the principles that humans are part of ecosystems and that we have a responsibility, in Leopold's words, to "preserve the integrity, stability, and beauty of the biotic community." This ethic serves--and is served by--modern goals of ecosystem management and resource stewardship. Fishery professionals, who are asked increasingly often to help refit patterns of human activity to patterns of the landscape, will do well to foster this ethic within society at large.

Professional ethics are noble on paper and encourage nobility in behavior and thought. But few professions have institutions and mechanisms for judging and punishing ethical misbehavior. The fisheries profession does not. Most fisheries people encounter ethical conflicts during their careers, such as unearned coauthorships, falsified data, dishonest testimony, mistreatment of others, and abuses of natural resources. Problems that seem clear-cut in some situations are ambiguous in others. One's own conscience and the support of respected colleagues may be the only guides to appropriate action in each situation.

## CERTIFICATION BENEFITS

Certification in the American Fisheries Society is based on peer evaluation of academic achievement, publications, and relevant professional experience. It is a valuable mechanism for advancing the profession and recognizing achievements in fisheries science. It establishes a benchmark of qualification and accomplishment that is recognized among society members, by courts of law, and increasingly by employers seeking highly qualified professional candidates.

Ecosystems and populations are very sensitive to manipulation by humans, including natural resource managers, and the need to measure and assure professional competence is particularly acute for resource stewardship. Standardized testing of applicants to determine their qualifications for certification is a need recognized by many, as is a linkage of certification renewal to continuing education programs. All renewable resource professions face the challenge of having certification recognized as a requirement for employment in resource management. One day they may collaborate to develop one or more standardized certification credentials that can be administered by an independent body and will be nationally recognized in the United States and Canada.

A workable certification program must set strong standards, yet be flexible enough to accommodate varying degrees of experience and specialization within a diverse profession. Differences of opinion about certification--how and to whom it should apply, and to what standards--make it very difficult to change a broad-based program, as the AFS experience has shown. Efforts to improve certification will be worthwhile if, as a result, aquatic resources become better managed by scientific and humanistic principles and if ethical standards for resource stewardship become enforceably codified in management practices.